Coastal Saline Soil Rehabilitation and Utilization Based on Forestry Approaches in China

Jianfeng Zhang

Coastal Saline Soil Rehabilitation and Utilization Based on Forestry Approaches in China

Springer

Jianfeng Zhang
Institute of Subtropical Forestry
Chinese Academy of Forestry
Fuyang, People's Republic of China

ISBN 978-3-662-52413-8 ISBN 978-3-642-39915-2 (eBook)
DOI 10.1007/978-3-642-39915-2
Springer Heidelberg New York Dordrecht London

Printed on acid-free paper

Springer is part of Springer Science+Business Media (www.springer.com)

Foreword I

Recent dramatic examples of the improvement and utilization of saline soil have received a great deal of worldwide attention. The development of practical afforestation techniques to turn saline lands green has been especially important to foresters and other professionals in environmental science fields. Presently, the area of saline soil resources covers about 10 million hectares in China, with nearly 5.3 million hectares located in the four coastal provinces of Jiangsu, Shandong, Hebei and Liaoning. Therefore, it is clear that abundant ecological, economic and social benefits could be attained if these saline lands are remedied and developed sufficiently.

During the past two decades, Chinese scientists have made great advances in the fields of rehabilitation and utilization of coastal saline soils. In particular, research has indicated that, by vegetation restoration in saline land, soil properties can be enhanced, and consequently the degraded soils can be ameliorated, and the mount of arable land area increased. Furthermore, the establishment of coastal shelterbelts can help these areas withstand natural disasters such as typhoons, tsunamis and storm surges, and bring about remarkable ecological and economic benefits. For more than 20 years, Prof. Zhang Jianfeng, from the Institute of Subtropical Forestry at the Chinese Academy of Forestry, has conducted outstanding research on salinity-tolerance mechanisms of trees and saline soil rehabilitation technologies. He initiated a valuable integrated technology system for saline soil amelioration focusing on vegetation restoration and forestry utilization. He summarizes the integrated technology and principles in his book entitled *Coastal Saline Soil Rehabilitation and Utilization Based on Forestry Approaches in China*. In light of the current environmental situation in China, the publication of this book will provide scientific guidance for ecological remediation in coastal saline lands in the future. The book contributes to the effort of realizing a harmonious relationship between ecological construction and economic development.

Performing research on saline soil improvement and utilization is especially important in coastal regions, which are heavily populated and have less available land. Obviously, reserve land resources will be increased when saline soils are rehabilitated. Hence, I hope that many more forestry professionals and scientists will recognize the importance of this task and will continue to make contributions to saline soil improvement and develop new techniques oriented to ecological rehabilitation. No doubt it will have a profound effect on alleviating the current disparities between the increasing population and the decrease in available land resources. Moreover, it will encourage civilization to become more ecologically-minded while helping to develop the beautiful country of China.

Beijing
January 30, 2013

Zhao Shucong

Foreword II

Actually salinization is one of the most serious long-term threats to the sustainability of land and water resources in the world. Salinization results from the accumulation of free salts to an extent that causes severe degradation of vegetation and soils. In developing countries, arable lands decrease with population growth and corresponding land use pressure. Accordingly a great amount of forests, grasslands and wetlands was exploited to cultivate and intensively management, leading to damage of original vegetation and soil structure in regions where salt accumulation occurs. On the other hand, irrational human practices have increased soil salinity by changing the natural balance of the water cycle in landscape by allowing excess recharging of groundwater. This reduces suitability to plant growth and increases the potential for other forms of land degradation such as soil erosion and structural decline. It is estimated that saline land increases annually 1–1.5 million ha worldwide.

In China, the situation seems much more crucial. With population growth and socioeconomic development, urbanization speeds up, more and more cultivated lands having been transferred, which results from heavier burden on the current farming land. Hence it is vital to reclaim and remedy disturbed land, especially saline land in the country. The key to deal with the issue is conducting rational land use systems, restoring salt tolerant plant communities and improving soil structure. Intensive research revealed that trees can tolerant a certain level of salinity due to high avoidance and tolerance abilities. Therefore in the viewpoint of biological amelioration of salt-affected soil, planting trees in salt affected landscapes may contribute to restore land capability for biomass production and reducing unproductive water lasses of the vegetation. After trees are established in saline land, the water table could be declined through increasing discharge and decreasing recharge. Additionally total salt content could be decreased, especially in soil surface profile and topsoil as well as soil physical and chemical state could be ameliorated.

In view of the principles, Prof. Dr. Zhang Jianfeng for a long time is going for probing that how to utilize and exploit saline soil as a geo-resource. During his research he emphases on ecological amelioration based on forestry approach from the angle of vegetation restoration instead of physical methods. Based on this consideration the book was developed. The first four parts mainly introduce characteristics of salt-affected soil and its amelioration by trees, including the concepts of saline soils, characteristics and occurrence of salt alkaloids, the resources condition and distribution, as well as urban salinization and ecological remediation which is a new environmental problem occurring recently. Moreover, principles and practice of afforestation in saline soils is discussed as a general description, covering the mechanism of salinity-tolerance of plants and salt-resistant plants selecting and breeding, the ecological rehabilitation principle and related technologies, approaches to the improvement progress as well as agroforestry and its application of saline soils along with general afforestation techniques in China.

Meanwhile, the Yellow River Delta as a typical and specific coastal region and the problem of wetland degradation and water shortage is discussed, too. Furthermore, the issue of saline soil utilization for biomass production is touched upon. It is well-known that biomass and bioenergy is keen demanded, but the land resources are limited in the country. Hence developing salt-affected soils for biomass production is necessary and significant. The both are concerned with soil salinization, consequently they are reciprocal in cause.

Finally in the fifth part some cases such as salinity tolerance features of trees in different growing stages are separately described based on the experiments of seed germination, physiological property measurement under salt stress, in addition to some parameters concerned with salt resistance of forest stocks, and black locust forests as biomass energy resource. Meanwhile planting techniques of *Tamarix chinensis* and *Nitraria sibirica* in worst salt-affected soil and its effect on soil amelioration were introduced. The disciplinary structure outlined above is exploited in some examples.

I am glad to read the manuscript and congratulate Prof. Zhang gaining a new achievement. Frankly spoken, this is the one of the best book I've reviewed on saline soil ameliorations based on forestry approach. It will provide an excellent resource for researchers, students, practical workers and other users. I believe that its value will emerge with years to come.

Dresden University of Technology
Dresden, Germany
March 10, 2013

Franz Makeschin

Preface

Across the world, soil salinization is becoming a serious environmental problem. It is an issue especially relevant in developing countries owing to their rapidly increasing levels of population growth and urbanization. As a result of these factors, there is increasing demand for staple goods such as grain and vegetables, which then leads to greater intensities of farming and corresponding adverse environmental impacts. One such impact is soil degradation, in which salinization, deforestation and desertification play major roles. In Australia, salinization is called "white death" because of its enormous harmful effects. Globally, an area of about 0.95 billion ha of soil is affected by salt, of which almost one tenth is distributed in China—mainly in the Huang-Huai-Hai, "three-river plain", and northwestern semi-arid and arid areas, as well as the coastal beach lands in the east of the country.

The coastline of China extends for more than 18,000 km, along which lay the Bohai Sea, Yellow Sea, East China Sea and South China Sea. With socioeconomic development, more and more people move to the east of the country to seek better opportunities and higher standards of living. No better example of the region's prosperity exists than China's "oriental pearl": Shanghai. Situated in the coastal area along the Yangtze River and the Yangtze River Delta, Shanghai represents an important component of the unique geographical advantage of this region, strategically positioned on land where resources are both rich and abundant. Today, coastal Shanghai plays a major role in expediting the development of the Yangtze River Delta region for industrial expansion. In doing so, China can continue to strengthen the productivity of its coastal areas nationally, as well as promote the development of the central and western regions of the country. As a result, exchange and cooperation between China and both Central/Northeast Asian countries and Europe can be maximized.

In view of the environmental situation faced currently by China, saline soils are important land resources and should play a crucial role in promoting socioeconomic development. The key to this is improving our understanding of how to recognize soil salinization, as well as how to manage and cope with it. The Chinese government has stated that the nation must provide 1.8 billion Mu (1 Mu = 1/15 ha) of cultivated land, and that the country must not fall below this "red line" in order

Acknowledgements

I take this opportunity to express my gratitude to the people who have been instrumental in the successful completion of this book.

First and foremost, I would like to thank Mr. Zhao Shucong (Director of the Chinese State Forestry Administration) and Prof. Franz Makeschin (former Dean of the Forestry Department, Technological University of Dresden and President of the German Society of Soil Science) for writing the Foreword of the book and giving me their valuable attention and time. Together, they were a great inspiration to me in working hard to complete the project, and I am truly grateful for their willingness to motivate me throughout.

Apart from the efforts of oneself, the success of any project depends largely on the encouragement and guidance of many others. The present book was no different in this regard, with its successful completion being very much the result of teamwork. Colleagues from the Institute of Subtropical Forestry, Chinese Academy of Forestry, such as Dr. Chen Guangcai, Dr. Shan Qihua, Prof. Jiang Jingmin, Prof. Yu Mukui (Project No. 2009BADB2B0304), Prof. Wang Yangdong, Prof. Zhou Benzhi, Prof. Gu Xiaoping, Dr. Wu Ming, Dr. Yang Xiaosheng, Dr. Li Sheng and Dr. Fang Xuezhi, all lent helping hands in various ways, as did my students Wang Qingbing, Li Zebo, Wang Ying, Wang Li, Wu Hao, Zhang Ying, Yang Quanquan, Li Jie and Sun Hui.

In addition, I take immense pleasure in thanking the Shandong Academy of Forestry, where I have worked for over 15 years. Colleagues such as Prof. Xing Shangjun, Prof. Jiang Yuezhong, Prof. Xu Jingwei, Prof. Liu Dexi, Prof. Song Yumin, Prof. Fang Yifu, Prof. Li Shanwen, Prof. Ma Hailin, Prof. Qin Guanghua, Dr. Du Zhenyu and Associate Prof. Ma Bingyao provide me with particular ongoing assistance. Furthermore, I also wish to thank my close friends Prof. Xi Jinbiao, Prof. Zhang Deshun, Prof. Li Xiufen and Prof. Qiao Yongjin, who have now left the Academy, for their kind co-operation and encouragement.

Time flies like a shuttle. In 1983, I enrolled at the Department of Forestry of Shandong Agricultural University. Since then, I have been learning and working in this field for 30 years, during which time I have gained a broad level of experience in forestry and related disciplines. I spent 7 years studying for a B.Sc. in forestry and

M.Sc. in silviculture. When I pursued my master's degree, Prof. Long Zhuangru—who sadly passed away in 2012—and Prof. Liang Yutang were my supervisors. These two gentlemen were rigorous scholars, and possessed a great accumulation of knowledge between them. However, they were at the same time both peaceful and humble in character, and I feel it is through their guidance and influence that I entered the world of science, and followed the road of academic research without looking back. I am enormously grateful for their inspiration, counseling and support.

After graduation in 1990, I began working at the Shandong Academy of Forestry, guided by Prof. Gong Hongzhu, who has long been working in the area of improvements of saline-alkali soils based on forest technology. He presided over the compilation of China's first textbook on the topic, entitled "Silviculture in Saline-alkali Land", which was a key milestone for saline-alkali soil amelioration and forest cultivation. It is through his advice and encouragement that I have enjoyed success along the rugged road of saline soil amelioration research.

Owing to the policy of reform and opening up in China, I also had the opportunity to study abroad between 1995 and 1997, sponsored by DAAD (the German Academic Exchange Service), whereupon I studied for an M.Sc. at the Institute of International Forestry and Forest Products at the Technological University of Dresden (TUD). Furthermore, I was later able to take up a position as a visiting scholar at the Institute of Soil Science, again at the TUD, from 2001 to 2002, supported by the China Scholarship Council. During these periods, many people provided me with an enormous level of help and advice; in particular, Prof. Franz Makeschin, Prof. Juergen Pretzsch, Prof. Holm Uibrig, Prof. Andreas Roloff, Dr. H. Pohris, Prof. Karl-Heinz Feger, Dr. Zhong Zheke, Dr. Dirk Landgraf, Dr. Frank Haubrig, Dr. Abi, and many others.

From 1999 to 2003 I studied for a Ph.D. at the College of Resource and Environment, Beijing Forestry University, supervised by Prof. Li Jiyue, who was the first person in China to be awarded a doctorate in silviculture (his Ph.D. tutor was Prof. Shen Guofang, academician and former president of the Beijing Forestry University). Prof. Li has a profound knowledge and rich experience in teaching. During the 4 years spent with Prof. Li, not only did I improve my level of professional knowledge, but he also imparted upon me the spirit of rigorous scholarship, a down-to-earth style of work, and the ability to rise above the crowd with personality and charm.

Having achieved my Ph.D., during 2003–2007 I then went on to perform postdoctoral research at the Institute of Forestry, Chinese Academy of Forestry, under the guidance of Prof. Peng Zhenhua, Chief Scientist and Kandidat Nauk from the former Soviet Union. Hence, he is both Chinese and Western in terms of his educational background, and possesses a rich farsightedness in scientific research, a pioneering scientific attitude, takes leave of dedication, and deeply affects everyone with whom he works. During the period, I also received support from Dr. Sun Qixiang, Prof. Zhang Xudong, Prof. Sun Zhenyuan, Prof. Meng Ping, Prof. Zhang Jianguo, Prof. Lu Mengzhu, Prof. Wang Cheng, Prof. Li Lubin, Prof. Wang Yan, Prof. Zhou Jinxing, Dr. Qiu Erfa, Prof. Sun Xiaomei, Prof. Wang Bing, Prof. Cui Lijuan, Prof. Fan Baomin, Prof. Guo Zhihua and Dr. Zhang Weiyin.

Finally, returning to the formulation of the book itself, I extend many thanks to Dr. Fan Lisa and Dr. Pei Xiaoli from Springer Asia for their kind help with manuscript preparation.

In short, I am hugely grateful to all my teachers, colleagues, other individuals and organizations (e.g., the Zhejiang Key Scientific and Technological Innovation Team) mentioned above, as well as my parents and friends, to whom I extend my sincere thanks as well. I would especially like to express my love and gratitude to my beloved family, my wife Mrs. Cao Junqing, and 14-year-old son Zhang Kai, for their understanding and endless love throughout the duration of my studies and work.

Hangzhou
May 9, 2013

Jianfeng Zhang

Contents

Part I Characteristics of Salt-Affected Soil and Its Amelioration by Trees

1 Concepts Concerned with Salt-Affected Soils 3
1.1 Concepts Concerned with Salt-Affected Soils 4
1.2 Process of Soil Salinization 5
1.3 Sources of Salts 5
1.4 Generic Properties of Saline Soils 6
References 7

2 Salt-Affected Soil Resources in China 9
2.1 Global Distribution of Salt-Affected Soils 10
2.2 Saline Soil Distribution in China 11
References 13

3 Characteristics of Salt-Affected Soil and Its Amelioration by Trees .. 15
3.1 Characteristics of Salt-Affected Soil 16
3.2 Functions of Trees in the Reclamation of Salt-Affected Land 17
3.2.1 Decreasing Recharge 18
3.2.2 Increasing Discharge 18
3.3 Saline Soil Remediation by Trees 19
References 21

4 Causes and Countermeasures of Urban Salinization 23
4.1 Process of Urban Salinization 24
4.2 Impacts of Salinity on Plant Growth and Urban Structure 26
4.3 Strategies for Dealing with Saline Soils 27
References 29

Part II Principles and Practice of Afforestation in Saline Soils

5 Salinity Tolerance and Tree Planting Techniques 33
5.1 Salinity Tolerance of Trees 34
5.2 Principles and Techniques of Afforestation in Saline Soils 34
5.2.1 Tree Species Selection 35
5.2.2 Improvement of Sites to Reduce Salt Content 36
5.2.3 Seedling Nursing 37
5.2.4 Planting 37
References 39

6 Agroforestry Systems for Combating Coastal Saline Soil 41
6.1 The Significance of Developing Agroforestry 42
6.2 Eastern China: The Main Environmental Problems 44
6.3 Salinity Sources and the Salinization Process in Eastern China ... 44
6.4 Effects of Agroforestry on Ameliorating Saline Soil 47
6.5 Agroforestry Systems in Eastern China 49
6.5.1 Selecting Tree Species 49
6.5.2 Shelterbelts in or Around Farmlands 50
6.5.3 Intercropping Agricultural Crops with *Ziziphus jujuba* 51
References 53

7 Strategies for Reclaiming and Ameliorating Saline Soil in the Yellow River Delta Region 55
7.1 Location and Natural Conditions of the YRD 56
7.2 Salinization Occurrence and Soil Features 58
7.3 Effects of Salt on Plant Growth 58
7.4 Strategies for Reclaiming and Ameliorating Saline Soil 59
7.4.1 Managing the Recharge and Discharge Areas 59
7.4.2 Drainage Practice and Leaching 59
7.4.3 Performance of Tillage 60
7.4.4 Applying Chemical Amendments 60
7.4.5 Adding Manure and Other Fertilizers 60
7.4.6 Planting Crops and Forage 61
7.4.7 Establishing Forests 62
References 64

Part III Saline Soil Utilization for Biomass Production

8 Biomass Production in Saline Soil 67
8.1 Saline Soil Resources in China 68
8.2 Flora Resources in Saline Soil for Biomass Production 70
8.3 Strategies for Reclaiming and Developing Saline Soil 71
References 72

9 Potential and Availability of Forest Biomass Energy Production in the East China Sea Regions 75
9.1 Importance of Developing the East China Sea Region 76
9.2 Main Constraints in the Region 76
9.3 Environmental Values of Coastal Areas 78
9.4 Economic Values of Coastal Regions 78
9.5 The Necessity and Feasibility of Developing Bioenergy 79
9.6 Functions of Energy Plants 80
References 82

Part IV Wetland Degradation and Water Shortage in the Yellow River Delta Region

10 Analysis of the Factors Driving Wetland Degradation 85
10.1 General Description of the Location 86
10.2 Wetland Types and Their Spatial Features 87
10.3 Causes of Wetland Degradation 89
10.3.1 Insufficient Water Replenishment 89
10.3.2 Serious Soil Salinization 89
10.3.3 Irrational Land Use 89
10.3.4 Pollution 89
10.4 Ecological Restoration Techniques for Degraded Wetlands 90
10.4.1 Ensuring Water Supply 90
10.4.2 Protecting the Virgin Plant Community 90
10.4.3 Assisting the Regeneration of Virgin Flora 90
10.4.4 Regulating Land Use 91
References 91

11 Impacts of Water Shortage 93
11.1 Basic Conditions of the Yellow River 93
11.2 Major Environmental Bottleneck 96
11.3 Impact of Climate Change on Runoff in the Yellow River Delta Region 97
11.4 Characteristics of the Water Resources 99
11.5 Current Statuses of Water Supply and Use 100
11.6 Main Questions Relating to Water Exploitation and Sustainable Use 100
11.6.1 Discrepancy Between Water Supply and Demand 100
11.6.2 Water Pollution 101
11.6.3 Water Wastage 101
11.6.4 Over-Exploitation of Groundwater 102
11.7 Countermeasures for Sustainable Use of Water 102
11.7.1 Formulating a Comprehensive Governance Plan 102
11.7.2 Reducing Water Wastage 103

11.7.3 Ensuring Basic Ecological Water Supply 103
11.7.4 Preventing Water Pollution 104
References 105

Part V Case Study

12 Establishment of *Tamarix chinensis* Stands and Its Amelioration of Saline Soil 109
12.1 Introduction 110
12.2 Materials and Methods 110
12.3 Results and Analysis 111
12.4 Conclusions and Discussion 114
References 115

13 Effects of *Nitraria sibirica* on the Improvement of Salt-Affected Soil 117
13.1 Introduction 118
13.2 Materials and Methods 119
13.3 Results and Analysis 119
13.4 Discussion and Conclusion 122
References 123

14 Tree Species Selection for Soil Salinity Resistance 125
14.1 Materials and Methods 126
14.1.1 Site Description 126
14.1.2 Plant Materials 126
14.1.3 Sampling and Analysis Methods 126
14.2 Results and Analysis 128
14.3 Conclusions 132
References 132

15 Effect of Tree Planting on Preventing Riverbank Soil Erosion 135
15.1 Introduction 136
15.2 Materials and Methods 136
15.2.1 Site Description 136
15.2.2 Plant Materials 137
15.2.3 Sampling and Analysis 138
15.3 Results 138
15.3.1 Tree Growth 138
15.3.2 Effects of Planted Seedlings on Soil Properties 139
15.4 Discussion 141
15.5 Conclusions 141
References 142

16 Function of Black Locust Plantation on Enhancing Soil Properties 145
16.1 Introduction 146
16.2 Methods 147

16.2.1 Study Area and Site Descriptions 147
16.2.2 Experimental Details 148
16.3 Results 149
16.3.1 Soil Salinity 149
16.3.2 Soil Na^+ Content and pH 150
16.3.3 Soil Nutrition 151
16.3.4 Soil Enzyme Activity 152
16.4 Discussion 153
16.5 Conclusions 154
References 154

17 Planting Black Locust (*Robinia pseudoacacia*) Forest as a Biomass Energy Resource 157
17.1 Introduction 157
17.2 Description of Natural Conditions in the Region 159
17.3 Materials and Methods 160
17.4 Results and Analysis 161
17.4.1 Standard Stock Biomass 161
17.4.2 Current Annual Biomass Increment and Net Annual Biomass Increment of the Tree 162
References 164

18 Responses of Plant Community Change on Wetland Degradation in Yellow River Delta Region 165
18.1 Introduction 166
18.2 Methods 167
18.2.1 Test Area Description 167
18.2.2 Sites Setting 167
18.2.3 Field Survey and Diversity Analysis 168
18.2.4 Soil Samples Collection and Pretreatment 168
18.2.5 Determination of Soil Properties 169
18.2.6 Determination of Soil Nutrients 169
18.2.7 Determination of Soil Enzyme Activity 169
18.3 Results 170
18.3.1 Flora Features at Different Successional Stages of Wetlands 170
18.3.2 Life Form Structures in Different Successional Stages 170
18.3.3 Species Composition and Plant Diversity at Different Successional Stages 171
18.3.4 Sequence of Vegetation Change During Wetland Degradation 172
18.3.5 Wetland Soil Characteristics at Different Succession Stages 173

18.3.6 Soil Physical Properties at Different Successional Stages of Wetlands 175
18.3.7 Soil Enzyme Activity at Different Successional Stages of Wetlands 175
18.4 Discussion and Conclusions 177
References 178

Index 181

Part I
Characteristics of Salt-Affected Soil and Its Amelioration by Trees

Chapter 1
Concepts Concerned with Salt-Affected Soils

Abstract With world population growth, the demand for food is increasing, but the area of arable land is reaching its limit. Moreover, these lands may even decrease owing to land-use changes associated with socioeconomic development. In China especially, which is one of the most populated countries in the world, its cultivated land area cannot keep pace with the population, and farmland is gradually declining as urbanization continues to progress. In order to meet the demand for grain production and alleviate the pressure on current arable land, saline land must be reclaimed. There are different opinions on how to define salt-affected soil in the field. Generally speaking, salt-affected soil can be defined as soil containing enough soluble salts to interfere with the growth of most crop plants. Also, the conductivity of the saturation extract tends to be greater than 4 mmhos/cm, the exchangeable-sodium percentage is less than 15 %, and the pH is usually less than 8.5. In order to describe specific cases, there are other terms concerned with salt-affected soils, and these are also discussed.

Keywords Soil degradation • Salt-affected soil • Soil salinization

With world population growth, the demand for food is increasing, but the area of arable land is reaching its limit. Moreover, these lands may even decrease owing to land-use changes associated with socioeconomic development. In China especially, which is one of the most populated countries in the world, its cultivated land area cannot keep pace with the population, and farmland is gradually declining as urbanization continues to progress. In order to meet the demand for grain production and alleviate the pressure on current arable land, wasteland must be reclaimed (Zhang 2004a, b). In China, there is a large area of salt-affected land, for example, in Lop Nur and Turfan (both around 5,000 km^2) and in Qaidam basin (20,000 km^2) (Gong 1986). Owing to the adverse conditions of such soils, crops are unable to grow, but planting trees by suitable means is possible. Forestation is not only

J. Zhang, *Coastal Saline Soil Rehabilitation and Utilization Based on Forestry Approaches in China*, DOI 10.1007/978-3-642-39915-2_1,

one way for saline–alkali soil utilization but can also improve soil properties and maintain ecological balance (Bohnert and Nelson 1995). Thus, it is relevant to discuss the characteristics of salt-affected soil and its amelioration by the planting and presence of trees.

1.1 Concepts Concerned with Salt-Affected Soils

There are different opinions on how to define salt-affected soil in the field. Generally speaking, salt-affected soil can be defined as soil containing enough soluble salts to interfere with the growth of most crop plants, as shown in Fig. 1.1. Also, the conductivity of the saturation extract tends to be greater than 4 mmhos/cm, the exchangeable-sodium percentage is less than 15 %, and the pH is usually less than 8.5 (Qureshi and Barrett-Lennard 1998). In order to describe specific cases, there are other terms concerned with salt-affected soils, and these are as follows:

1. Alkaline soil
2. Any soil that has a pH greater than 7.0
3. Sodic soil
   (a) A soil containing sufficient sodium to interfere with the growth of most crop plants
   (b) A soil having an exchangeable-sodium percentage of 15 % or more

Fig. 1.1 Landscape of salt-affected soil (Photograph taken by Xi Jinbiao)

4. Saline–alkali soil
   (a) A soil containing enough exchangeable sodium to interfere with the growth of most crop plants and containing appreciable quantities of soluble salts. The exchangeable-sodium percentage is greater than 15 %, the conductivity of the saturation extract is greater than 4 mmhos/cm at 25 °C, and the pH is usually 8.5 or less in saturated soil (Sharma 1986).
   (b) A saline–alkali soil has a combination of harmful quantities of salts and either a high alkalinity or high content of exchangeable sodium, or both, distributed throughout the profile in such a way that the growth of most crop plants is reduced. This type of soil can also be referred to as saline–sodic soil.

1.2 Process of Soil Salinization

There are two kinds of soil salinity: dryland salinity (occurring on land not subject to irrigation) and irrigated-land salinity. Both describe areas where soils contain high levels of salt. Usually, plants and soil organisms are killed or their productivity is severely limited on salt-affected lands (Marcar et al. 1999).

Irrigated-land salinity arises because of the clearance of trees and replacement with shallow-rooted crops, overwatering, seepage from irrigation channels, and impaired natural drainage. The main cause of high water tables and dryland salinity is land-use change, particularly the broadscale replacement of deep-rooted native vegetation such as trees and perennial native grasses with shallow-rooted annual agricultural crops and pastures that use less water.

In recent years, it has been recognized that many other land management practices aside from irrigation and clearing for agriculture may contribute to rising water tables and saline seepage in many areas. Such practices include excessive watering of lawns and gardens, tree clearing for farming and/or urban development, irrigation of public recreation areas, poorly designed septic and sewerage systems, and the development of roads, bridges, railway, and flood control banks, which interfere with natural surface drainage patterns and induce ponding of surface water (Makeschin et al. 2001).

1.3 Sources of Salts

Saline waters (2–3 g/L) and salty waters (>3 g/L) occur in the groundwater systems that underlie most of the region (7,230 km^2 or 51 %). There are high levels of Cl^-, SO_4^{2-}, and Na^+, with the total salt content depending directly on the concentrations of these ions. Groundwater is the most active cause of saline soils. However, saline soils have gradually decreased in extent because groundwater levels have declined during the past 20 years.

The dominant source is a shallow layer of salty groundwater. Water from upland river systems played a major role in the past, but its importance has decreased in the last 30 years and such water is no longer a source of salt.

Wind and air are other sources of salt accumulation in coastal zones. Salt from the sea can be carried long distances inland by wind as spindrift or can fall to the ground with rain after being carried by warm northwesterly winds. "Desalted Chao soil" results from large-scale improvements in saline soils. Surface soils have gradually lost salt as water, fertilizer, and salt regimes have changed. However, salt remains below the subsoil and is therefore a potential source of surface salt, which could accumulate if groundwater levels rise.

1.4 Generic Properties of Saline Soils

Saline soils are soils that have salts that occur in amounts harmful to germination of seeds and plant growth. In Australia, it is called "white death" (see Fig. 1.2).

Saline soils often exhibit a whitish surface crust when dry. Soil salts found in China are of three types: (i) chlorides (Cl^-), (ii) sulfates (SO_4^{2-}), and (iii) carbonates (CO_3^{2-}). Most saline soils in coastal regions have high amounts of chloride salts. However, in some parts of inland regions, extensive areas of saline soils are also composed of sulfate salts. Saline soils with high amounts of carbonates of sodium (sodium bicarbonate) may also occur and are usually associated with

Fig. 1.2 Whitish surface crusts in saline soil (Photograph taken by Xi Jinbiao)

coarse textured soils. The solubility of gypsum (calcium sulfate, $CaSO_4$) is commonly used as the standard for comparing the solubility of salts. Consequently, salts more soluble than gypsum are considered to be soluble and cause salinity, such as sodium chloride (table salt, NaCl) and sodium sulfate (thenardite, Na_2SO_4). In some soils, sulfate-rich (e.g., saline acid sulfate soils) or boron-rich salts can be found in excessive or toxic amounts for plant growth. Another type of soil problem occurs when sodium levels are high in relation to calcium and magnesium in the soil. These soils are very sticky and slippery when wet and very hard, cloddy, and prone to crusting when dry.

In the worst affected areas, soluble salts are precipitated at the surface, primary clay minerals are destroyed, and the soil becomes susceptible to erosion.

It is well known that China is one of the most populated countries in the world, but its cultivated land area cannot keep pace with the population, and farmland is gradually declining as urbanization continues to progress (Sun and Zhang 2006). In order to meet the demand for grain production and alleviate the pressure on current arable land, saline land must be reclaimed. Forestation is not only one way for saline–alkali soil utilization but also can improve soil property and maintain ecological balance.

Salt-affected soils are distributed extensively across the Earth. Although the causes are various, in general terms, such soils result from the accumulation of free salts to such an extent that degradation of the soil and vegetation occurs. Meanwhile, unwise human practices have also increased soil salinity in some areas by changing the natural balance of the water cycle in the landscape by allowing excess recharging of groundwater. This reduces suitability for plant growth and increases the potential for other forms of land degradation, such as soil erosion and structural decline. Thus, salinization is one of the most serious long-term threats to the sustainability of land and water resources in the world. Trees can tolerate a certain level of salinity, thanks to evolution and adaptation. Therefore, in terms of the biological amelioration of salt-affected soil, planting trees is necessary and significant. After trees are established on saline land, the water table can decline through increased discharge and decreased recharge. Additionally, total salt content can decrease—especially in the soil surface profile and topsoil—and adverse soil physical and chemical states can be ameliorated.

References

Bohnert HJ, Nelson DE (1995) Adaptation to environmental stresses. Plant Cell 7:1099–1111

Gong HZ (1986) Silviculture in saline soil. China Forestry Press, Beijing, pp 62–66

Makeschin F, Landgraf D, Zhong Z (2001) Nachhaltige Agroforstwitschaft auf salinen Koestenboeden der Provinz Zhejiang (VR China). Wissenschaftliche Zeitschrift der Technischen Universität Dresden 4:102–108

Marcar N, Ismail S, Hossain A (1999) Trees, shrubs and grasses for saltlands. Australian Centre for International Agricultural Research, Canberra, pp 26–89

Qureshi RH, Barrett-Lennard EG (1998) Saline agriculture for irrigated land in Pakistan: a handbook. Australian Centre for International Agricultural Research, Canberra, pp 84–125

Sharma SK (1986) Saline environment and plant growth. Kluwer Academic Publishers, Netherland, pp 128–193

Sun QX, Zhang JF (2006) Discussion on forestry sustainable development in the Yellow River delta region. Chin Forest Sci Technol 5(4):63–67

Zhang JF (2004a) Agroforestry and its application in amelioration of saline soils in eastern China coastal region. Forest Study China 6(2):27–33

Zhang JF (2004b) Principles and practice of forestation in saline soil in China. Chin Forest Sci Technol 3(2):62–70

Chapter 2
Salt-Affected Soil Resources in China

Abstract In developing countries, arable lands are decreasing as populations increase. Accordingly, a great amount of forests, grasslands, and wetlands are being exploited through intensive cultivation and management. This leads to damage of the original vegetation and soil structure and causes soil salinization. It is estimated that the total area of saline land is increasing across the globe by 1–1.5 million ha every year. In China, where the causes and characteristics of saline land can be divided into five regions (as will be explained in this chapter), there is little doubt that the current situation regarding salt-affected land is highly critical. The country has some of the highest populated lands in the world, and as population continues to grow, accompanied by ongoing socioeconomic development and increasing levels of urbanization, more and more cultivated lands are being transformed into other uses, placing an even heavier burden on the remaining farmland. Meanwhile, soil degradation is worsening due to anthropogenic environmental impacts and intensive management. Thus, it is vital to reclaim and remedy affected land—especially salt-affected land—throughout the country. The key to dealing with this issue is to operate a sensible land-use system, restore plant communities, and improve soil structure.

Keywords China • Soil resources • Saline soil

Science and technology in the twenty-first century have reached a new level, helping to boost socioeconomic development. However, there are many serious environmental problems faced by humankind associated with this development, such as climate change, contamination and pollution, soil desertification, and salinization (Makeschin et al. 2001; Mainguet 1999; Zhang 2002). In developing countries, arable lands decrease as populations increase. Accordingly, a great amount of forests, grasslands, and wetlands are being exploited through cultivation and intensive management, leading to damage of the original vegetation and soil structure, and ultimately soil salinization (Malcolm and Sumner 1998; Marcar et al. 1999;

J. Zhang, *Coastal Saline Soil Rehabilitation and Utilization Based on Forestry Approaches in China*, DOI 10.1007/978-3-642-39915-2_2,

Zhang and Li 2002). It is estimated that the total area of saline land is increasing across the globe by 1–1.5 million ha every year (Kovda 1983). In China, the situation is even more critical. The country has some of the highest populated lands in the world, and as the population continues to grow, accompanied by ongoing socioeconomic development and increasing levels of urbanization, more and more cultivated lands are being transformed into other uses, placing an even heavier burden on the remaining farmland. Meanwhile, soil degradation is worsening due to anthropogenic environmental impacts and intensive management. Thus, it is vital to reclaim and remedy affected land—especially salt-affected land—throughout the country. The key to dealing with this issue is to operate a sensible land-use system, restore plant communities, and improve soil structure.

2.1 Global Distribution of Salt-Affected Soils

Salt-affected soils are distributed widely across the Earth, from cold and temperate zones to subtropical and tropical zones. The distribution is shown in Tables 2.1 and 2.2 (Szabolcs 1989; Gupta et al. 1987).

Table 2.1 Global distribution of salt-affected soils

Region	Area (1,000 ha)	Percentage (%)
North America	15,755	1.65
Mexico and Central America	1,965	0.21
South America	129,163	13.53
Africa	80,538	8.43
South Asia	87,608	9.17
North and Central Asia	211,686	22.17
Southeast Asia	19,983	2.09
Australasia	357,330	37.42
Europe	50,804	5.32
Total	954,832	

Table 2.2 Distribution of saline soils in the most significantly affected countries and regions

Country/region	Area (1,000 ha)
Australia	357,240
Former USSR	170,720
China	36,658
Indonesia	13,213
Pakistan	10,456
India	7,000
Iran	6,726
Saudi Arabia	6,002
Mongolia	4,070
Malaysia	3,040

2.2 Saline Soil Distribution in China

The distribution of saline soil in China varies greatly because of the country's huge land area and various climates. Not only is there variation in area but also in terms of the salt ingredients and salinization processes involved. To illustrate this variation, China can be divided into five regions based on salinity characteristics and their causes.

1. *Northwest inland saline land.* This region covers most of Xinjiang, Chaidamu Basin in Qinghai, the Hexi Corridor in Gansu, and west Inner Mongolia. The area is characterized by a terrestrial climate, annual precipitation of 100–300 mm, and a water table of 3–10 m (1–2 m in some areas). The groundwater mineral content is 3–5 g/L, with a maximum of 10 g/L. The major anions present are Cl^- and $SO_4{}^{2-}$, and the total soil salinity is 1–4 % in the upper 20 % of the topsoil.
2. *Semiarid saline land in the middle reaches of the Yellow River.* This area includes east Qinghai and Gansu, Hetao District in Ningxia and Inner Mongolia, as well as the valley plain in Shaanxi and Shanxi. It is characterized by complex landforms, poor drainage facilities, and drought. Annual precipitation is 150–400 mm, with high evapotranspiration due to wind. Saline soil is distributed as strips in the Yellow River alluvial valley and Loess Plateau.
3. *Semiarid and arid lowland saline lands in the Yellow–Huai–Hai River Plain.* This land is situated from the lower reaches of the Yellow–Huai–Hai River Plain to the North China Plain, covering Beijing, Tianjin, Hebei, Shandong, Henan, Anhui, and Jiangsu provinces. Annual precipitation increases gradually from 400 mm in the north to 800 mm in the south. The water table is 1–3 m, and the groundwater mineral content is 2–5 g/L, with a maximum of 10 g/L. The major anions present are Cl^-, $SO_4{}^{2-}$, and $CO_3{}^{2-}$.
4. *Semiarid and semi-humid lowland saline land in Northeast China.* This region includes the Songhua–Nenjiang River Plain, the west of Liaoning Basin, the three-river (i.e., Songhua River, Heilongjiang River, Wusuli River) plain, and Hulenber District. Annual precipitation is 500–700 mm, and the water table is 3 m. The groundwater mineral content is 2–5 g/L, with a maximum of 10 g/L, and the major anions present are $CO_3{}^{2-}$ and $HCO_3{}^-$. The soil salinity is 0.3 %.
5. *Coastal semi-humid saline land.* These lands cover coastal regions in eastern and southern China, as well as the northern Yangtze River, which are characterized by a monsoon climate. Annual precipitation increases gradually from 600 mm in the north to 1,000 mm in the south. The water table is 0.5–2.5 m, with a groundwater mineral concentration of over 10 g/L (maximum of 50 g/L). The soil salinity is usually more than 0.4 %, and the main anion present is Cl^- (see Fig. 2.1).

With world population on the rise, the demand for food is increasing. However, the area of arable land cannot keep pace. In fact, it is actually *decreasing* because of land-use changes stemming from socioeconomic development. Especially in China, the most populated country in the world, the cultivated land area is far from sufficient in terms of population growth. Moreover, farmland is gradually declining with the rapid progress of urbanization.

Fig. 2.1 Saline soil in Dongying, Shandong province (Photograph taken by Zhang Jianfeng)

The improvement and use of saline soils should take into account seasonality, local conditions, and crop types and combine specialized use with comprehensive controls. Technical measures should be based on knowledge about the causes of the saline soil and the principles of movement of water and saltwater. It is important to find solutions for the permanent control of salinization as well as quick temporary solutions. In order to meet the demand for grain production, and alleviate the pressure on current arable land, wasteland must be reclaimed. Forestation is not only helpful for saline–alkali soil utilization, but it can also improve soil properties and maintain ecological balance (see Fig. 2.2).

Salt-affected soils are distributed widely across the Earth. Although there are various causes for the occurrence of salinization, in general terms, it results from the accumulation of free salts to such an extent that degradation of vegetation and soils takes place. Meanwhile, unwise human practices have also increased soil salinity by changing the natural balance of the water cycle in the landscape and allowing excess recharging of groundwater. This reduces suitability for plant growth and increases the potential for other forms of land degradation, such as soil erosion and structural decline. Thus, salinization is one of the most serious long-term threats to the sustainability of land and water resources in the world. Trees can tolerate a certain level of salinity, thanks to evolution and adaptation. Therefore, from the viewpoint of biological amelioration of salt-affected soil, planting trees is necessary and significant. After trees are established on saline land, the water table can decline through increased discharge and decreased recharge. Additionally, total salt content can decrease—especially in the soil surface profile and topsoil—and adverse soil physical and chemical states can be ameliorated.

Fig. 2.2 Plantation of *Casuarina equisetifolia* L. at Sanya, Hainan (Photograph taken by Zhang Jianfeng)

References

Gupta GN, Mohan S, Prasad KG (1987) Salt-tolerance of selected tree seedlings. J Trop Forest 3(3):217–227

Kovda VA (1983) Loss of productive land due to salinization. Ambio XII(2):91–93

Mainguet M (1999) Aridity drought and human development. Springer, Berlin

Makeschin F, Landgraf D, Zhong Z (2001) Nachhaltige Agroforstwitschaft auf salinen Koestenboeden der Provinz Zhejiang (VR China). Wissenschaftliche Zeitschrift der Technischen Universität Dresden 4:102–108

Malcolm E, Sumner RN (1998) Sodic soils-distribution, properties, management, and environmental consequences. Oxford University Press, New York

Marcar N, Ismail S, Hossain A (1999) Trees, shrubs and grasses for saltlands. Australian Centre for International Agricultural Research, Canberra, pp 26–89

Szabolcs I (1989) Salt-affected soils. CRC Press, Boca Raton

Zhang JF (2002) Research on saline soil amelioration and forestation techniques. J Northeast Forest Univ 30(6):124–129

Zhang JF, Li XF (2002) Strategies to reclaim and ameliorate saline soil in the Yellow River Delta Region. In: Chang C, Dobing B (eds) International conference on environmentally sustainable agriculture for dry areas for the 3rd millennium proceedings. Dobing Enterprises, Canada, pp 264–269

Chapter 3
Characteristics of Salt-Affected Soil and Its Amelioration by Trees

Abstract China is the most populated country in the world, but its cultivated land area is not large enough to fully provide for it. Farmland is gradually declining as a result of socioeconomic development and rapid urbanization. In order to meet the demand for grain production and alleviate the pressure on current arable land, wasteland must be reclaimed. Forestation is not only useful for saline–alkali soil utilization but can also improve soil properties and maintain ecological balance. Salt-affected soils are distributed widely across the earth; and, although there are various causes for the occurrence of salinization, generally it results from the accumulation of free salts to such an extent that degradation of vegetation and soils takes place. Furthermore, unwise human practices have also increased soil salinity by changing the natural balance of the water cycle in the landscape and allowing excess recharging of groundwater. This reduces suitability for plant growth and increases the potential for other forms of land degradation, such as soil erosion and structural decline. Thus, salinization is one of the most serious long-term threats to the sustainability of land and water resources in the world. Trees can tolerate a certain level of salinity, thanks to evolution and adaptation. Therefore, from the viewpoint of biological amelioration of salt-affected soil, planting trees is necessary and significant. After trees are established on saline lands, the water table may decline through increasing discharge and decreasing recharge. Additionally, the total salt content might decrease, especially surface soil layers and topsoil. The potential for the amelioration of soil physical and chemical states also exists.

Keywords Salinization • Remediation • Trees • Forestation • China

Land salinity is closely related to other land degradation processes, such as soil erosion, acidification, and structural decline. Aside from that, soil salinity usually leads to high pH. Soil fertility under a high pH varies because the availability of phosphorus and most micronutrients is reduced. This makes placement more

J. Zhang, *Coastal Saline Soil Rehabilitation and Utilization Based on Forestry Approaches in China*, DOI 10.1007/978-3-642-39915-2_3,

important, because urea losses to volatilization are greater and risk of injury from seed-placed urea is increased (Makeschin et al. 2002).

Soil structure in sodic soils is poor and permeability is low. If a sodic clay layer occurs near the surface of sodic soils, it often acts as a barrier to the roots. Hence, most roots are restricted to the topsoil above the claypan because the movement of water, nutrients, and gases in sodic B horizons is too slow. When dry, the B horizon can be so hard that it physically restricts root penetration. The overall effect on plant growth is one of stress, similar to that caused by extremely dry or saline conditions.

High soil salinity (high EC) causes dehydration of plant cells, which in turn results in reduced plant growth and often death. Dissolved salts cause dehydration of plant cells because the osmotic potential of soil water is decreased. Water will flow from the high osmotic potential (low salts in plant cells) to low osmotic potential (high salts in soils). Thus, plants simply cannot extract water from soils when soil solutions have lower osmotic potentials than plant cells. The effect on plants is similar to drought stress. Yields of many crops are reduced when the soil extract EC reaches 4 dS/m. Yields will decline proportionately as EC levels increase above 4 dS/m. Some crops, such as sugar beets and cotton, are tolerant to ECs of between 4 and 8 dS/m. The growth and yields of most crops are affected at EC values of greater than 16 dS/m.

3.1 Characteristics of Salt-Affected Soil

Soil salinity is a soil condition where water-soluble salts in the crop-rooting zone impede crop growth. The severity of the effects of salinity and strategies to address the problem depends on soil testing to identify the amount and type of salts present. The clearing of native vegetation is a major cause of all these problems. Soil acidity may be a key factor in increasing groundwater recharge and thus hastening the rate of a rise in the water table.

High salt content increases the osmotic potential of the soil solution and prevents the uptake of water by crops. Crops are generally most sensitive to salinity during germination and emergence, and some plants are more sensitive to salinity than others, depending on their growth characteristics and root system (Zhang 2002).

Another type of soil problem occurs when sodium levels are high in relation to calcium and magnesium in the soil. These soils are very sticky and slippery when wet and very hard, cloddy, and prone to crusting when dry.

In the worst affected areas, soluble salts are precipitated at the surface, primary clay minerals are destroyed, and the soil becomes susceptible to erosion. Grossly salt-affected land, where vegetation has been severely reduced or lost, is more prone to erosion, and the erosion of topsoil can then also hasten further expression of land salinity by exposing saline conditions under the soil. Once land is affected by more than one form of land degradation, preventive and remedial measures are much more difficult.

3.2 Functions of Trees in the Reclamation of Salt-Affected Land

As mentioned above, salts are a critical factor in hindering tree survival and growth in these lands. Extra salinity in soil is harmful to trees. On one level, it influences the health of roots, refraining or hindering the absorption of mineral elements and water; while on another level, salts can directly injure roots, damaging plant tissues and disturbing the normal physiological activity of trees (Lieth and Al-Masoon 1993).

Nevertheless, plants can develop different mechanisms to tolerate high salt levels, including the restriction of salt uptake, sequestration of salt in cells, and osmotic adjustment. Different trees have developed various ways to tolerate salt. Some, such as *Elaeagnus angustifolia*, use salt exclusion, refusing to take up salts (Zhang 2004a); while others, such as *Tamarix* spp., use salt secretion, which means that they can secrete salt through glands after salts have been absorbed from soil (see Fig. 3.1).

There are also trees that use salt dilution, meaning they can dilute salt in order to maintain lower levels of salt in cells, as is the case with *Hordeum vulgare*, while others use salt avoidance, including *Rhizophora apiculata*. Some trees are characterized by a comprehensive array of salt tolerance mechanisms; for example, *Fraxinus* spp. can endure salt stress through a number of complicated measures and has thus become an important pioneer tree species for planting in saline soil

Fig. 3.1 Trees planted in saline soil (Photograph taken by Zhang Jianfeng)

(Doran and Turnbull 1997). The traits of these trees are useful because they can help ameliorate saline soil by biological means based on the identification of the occurrence of soil salinization.

Groundwater recharge is the amount of water being added to the groundwater. If this is higher than discharge, which is the amount of water lost from the groundwater, then the water table rises. As it does, the water dissolves salt held in the soil profile, and the salt becomes more and more concentrated as the water moves upwards. If the salty water keeps rising, it eventually reaches the surface and subsurface layers of the soil. The water then evaporates, leaving the salt behind. Trees help control ground water levels in two ways: by decreasing recharge and by increasing discharge (Qureshi and Barrett-Lennard 1998).

3.2.1 Decreasing Recharge

Most groundwater recharge is supplied by rainfall (except in the case of irrigated-land salinity), and more of it reaches the groundwater when trees are cleared. This is because trees develop extensive root systems to trap the water, which is then used for tree growth or returned to the atmosphere by evaporation and transpiration. Scientists estimate that the amount of water that percolates below the root zone of crops and pastures can be 10–100 times the amount percolating below trees (Sun and Zhang 2006).

3.2.2 Increasing Discharge

Trees also play a role in discharging groundwater. The roots of many tree species reach deep underground, often making contact with the water table. During drought, such trees use this water in order to survive and keep growing. In contrast, annual crops that farmers plant on cleared land usually do not have deep roots (Zhang and Li 2002). Moreover, many crops only grow during winter, spring, and early summer, while most trees require water all year round.

Trees planted in salt-affected land could improve soil quality and enhance soil fertilization (see Fig. 3.2). Organic matter would be increased through leaf fall, and another factor is that roots move and modify the soil structure. Therefore, after saline land is forested, total salt content can be decreased, especially in the upper soil layers and topsoil. Additionally, soil physical and chemical states can be ameliorated (Staples 1984).

Soil salinization is usually caused by saline groundwater interaction. Other environmental factors (e.g., physiographical conditions, soil types, rainfall) and human activities (e.g., irrigation regimes and different socioeconomic settings) also have influences on the process. Therefore, there is no single technique or agricultural system that will be applicable to all areas and conditions.

Fig. 3.2 Soil property is enhanced by planting trees (Photograph taken by Zhang Jianfeng)

Trees play an important role in the improvement and reclamation of salt-affected soils. The complexity of the management of salt-affected soils requires a combination of agronomic and management practices. The process should involve carefully defining the main production constraints and requirements based on a detailed and comprehensive investigation of soil characteristics, such as water monitoring (rainfall, irrigation water, and the water table). A survey of local conditions is necessary, including climate and crops, as well as the economic, social, political, and cultural environment and existing farming systems practiced now and in the past.

3.3 Saline Soil Remediation by Trees

There are different ways to ameliorate saline soils. Successful remediation of highly saline–sodic, clayey soil via leaching requires the maintenance of some degree of hydraulic conductivity (HC) throughout the soil profile. It is generally thought that the degree of swelling in salt-affected soils can be manipulated by controlling EC and SAR. Ideally, SAR, or the ratio of sodium to calcium and magnesium cations, should be reduced prior to reductions in EC to prevent loss of HC caused by swelling and dispersion of soil clays.

A study was carried out to examine the efficacy of several treatment combinations intended to manipulate saline–sodic soils, chemically and physically, during

Table 3.1 Soil salt content inside and outside the forest area

Sample location	Soil depth (cm)	Salt content (%)				Mean
Inside the forest	0–20	1.10	0.73	1.17	1.17	1.06
	20–50	0.97	0.56	1.07	0.83	0.86
Outside the forest	0–20	2.17	0.90	2.40	1.54	1.75
	20–50	1.10	0.57	1.07	1.10	0.96

Table 3.2 Comparison of physical properties and salinity at different sites

Site type	Soil depth (cm)	Soil moisture (%)	Total porosity (%)	Soil capacity (g/cm^3)	Organic matter (%)	Total salt (%)	Mean (%)
Bare land	0–20	20.12	45.18	1.45	0.245 9	1.78	1.034
	20–40	28.39	46.42	1.42	0.111 2	1.00	
	40–60	21.63	46.79	1.41	0.044 8	0.71	
	60–100				0.078 5	0.84	
Woodland	0–20	31.60	53.91	1.20	0.325 0	0.59	
(*Tamarix*)	20–40	28.58	47.92	1.38	0.133 6	0.48	
	40–60	29.08	44.53	1.45	0.033 6	0.53	0.544
	60–100				0.100 9	0.56	

the early stages of remediation. Treatments included the application of calcium chloride, gypsum, and bagasse, alone and in various combinations. Additionally, subsurface drainage and sprinkler irrigation systems were installed over the entire site. Subsurface drains and collection sumps were installed to overcome the seasonally high water table (fall/winter), and drained waters collected in sumps were hauled to a nearby deep injection well for disposal. A sprinkler irrigation system was installed to supply water during the dry season (spring/summer) to take advantage of the seasonally low water table.

A high incidence of middle- and high-canopy shrubs and trees left or planted on farms serve as physical windbreaks, and control of the wind in this manner helps to reduce the rate of evapotranspiration from crops growing on the land. This is necessary particularly in those farm holdings located in drier areas. Indeed, this notion that with more trees there is less evaporation and less salt accumulation in topsoil has been tested, and examples of the functions of trees in this regard is demonstrated by the results presented in Tables 3.1 and 3.2 (Zhang 2004b).

Table 3.1 shows salinity inside the forest was reduced by 39 % in the 0–20-cm soil layer, while in the 20–50-cm layer, the reduction was 10 %. The effect of trees on the amelioration of saline soils was therefore apparent. Furthermore, from Table 3.2, it can be seen that a *Tamarix* forest can reduce the soil salt concentration effectively, especially in the 0–40-cm layer. Organic matter and soil moisture increased obviously, while soil capacity and soil total porosity were also enhanced.

Extensive site characterization led to the identification of sodium chloride as the primary salt contaminant in soils associated with oil production activities.

The deleterious effects of salt contamination, particularly NaCl, upon a soil's chemical and physical properties are well known. Toxicity to plants as well as the deterioration of soil physical properties is a common occurrence. This can be evident as reduced infiltration and aggregation ultimately enhance the potential for soil erosion. It should be noted that the degree to which sodicity affects any given soil is influenced by factors such as soil mineralogy, pH, texture, and organic content.

Within the framework of salt-affected soil improvement, particular attention is paid to the salinity tolerance of crop plants according to the specific stage of improvement (Zhang 2003). In the first stages of improvement, the most salt-tolerant plants, such as *Nitraria* spp., *Tamarix* spp., *Lycium chinense*, and *Amorpha fruticosa*, are planted. In the later stages, when soils become more fertile and have lower salt content, it is possible to grow crops more sensitive to salinity, such as rice, soybean, and corn (Crouch and Honeyman 1986; Song 2001).

Clearly, salinization is one of the most serious long-term threats to the sustainability of land and water resources in the world. However, trees can tolerate a certain level of salinity, thanks to evolution and adaptation. Therefore, in terms of the biological amelioration of salt-affected soil, planting trees is necessary and significant. After trees are established on saline land, the water table can decline through increasing discharge and decreasing recharge. Additionally, total salt content can be decreased, especially in the surface layers and topsoil. The physical and chemical state of the soil can also be ameliorated.

References

Crouch RJ, Honeyman MN (1986) The relative salt-tolerance of willow cuttings. J Soil Conserv 42(2):103–104

Doran JC, Turnbull JW (1997) Australian trees and shrubs: species for land rehabilitation and farm planting in the tropics. Australian Centre for International Agricultural Research, Canberra, pp 235–256

Lieth H, Al Masoon AA (1993) Towards the rational use of high salinity tolerant plant. Kluwer, Netherlands, pp 12–127

Makeschin F, Fischer R, Zhong Z (2002) Bodenchememische Eigenschaften saliner Koestenstandorten der Provinz Zhejiang (China). In: Makeschin F (ed) Mitteilgn Dtsch. Bodenkundl. Gesellsch/Springer, Berlin, pp 68–74

Qureshi RH, Barrett-Lennard EG (1998) Saline agriculture for irrigated land in Pakistan: a hand book. Australian Centre for International Agricultural Research, Canberra

Song YM (2001) Study on salt-tolerance and major planting techniques of *Tamarix chinensis*. In: Liu MY (ed) The proceedings of international symposium on halophyte utilization and regional sustainable development of agriculture. Meteorology press, Beijing, pp 68–72

Staples RC (1984) Salinity tolerance in plants strategies for crop improvement. Kluwer, Netherland, pp 124–129

Sun QX, Zhang JF (2006) Discussion on forestry sustainable development in the Yellow River delta region. Chin Forest Sci Technol 5(4):63–67

Zhang JF (2002) Research on saline soil amelioration and forestation techniques. J Northeast Forest Univ 30(6):124–129

Zhang JF (2003) Advances of research on the mechanism of plant salinity tolerance and breeding of salt-tolerant plants. World Forest Res 16(2):16–22

Zhang JF (2004a) Study on cultural technologies and salt-resistance of *Nitraria sibirica* in Coastal Areas with serious salt-affected soil. Chin Forest Sci Technol 3(4):12–16

Zhang JF (2004b) Agroforestry and its application in amelioration of saline soils in eastern China coastal region. Forest Study China 6(2):27–33

Zhang JF, Li XF (2002) Strategies to reclaim and ameliorate saline soil in the Yellow River Delta Region. In: Chang Chi, Brian D (eds) International conference on environmentally sustainable agriculture for dry areas for the 3rd millennium proceedings. Dobing Enterprises, Canada, pp 264–269

Chapter 4
Causes and Countermeasures of Urban Salinization

Abstract There are different ways to ameliorate saline soils in urban regions. Successful remediation of highly saline–sodic, clayey soil via leaching requires the maintenance of some degree of hydraulic conductivity (HC) throughout the soil profile. It is generally thought that the degree of swelling in salt-affected soils can be manipulated by controlling EC and SAR. Ideally, SAR, or the ratio of sodium to calcium and magnesium cations, should be reduced prior to reductions in EC to prevent loss of HC caused by swelling and dispersion of soil clays. A study was conducted to examine the efficacy of several treatment combinations intended to manipulate saline–sodic soils, chemically and physically, during the early stages of remediation. Treatments included calcium chloride, gypsum, and bagasse, alone and in various combinations. Additionally, subsurface drainage and sprinkler irrigation systems were installed over the entire site. Subsurface drains and collection sumps were installed to overcome the seasonally high water table (fall/winter), and drained waters collected in sumps were hauled to a nearby deep injection well for disposal. A sprinkler irrigation system was installed to supply water during the dry season (spring/summer) to take advantage of the seasonally low water table.

Keywords Urban salinization • Remediation • Tree species

Science and technology have reached a new stage in which socioeconomic development can take place rapidly, and one of the most important indicators of this is urbanization. However, a consequence of such development for the human population is the related environmental problems, such as climate change, contamination and pollution, and soil desertification and salinization, which are extremely serious issues being faced in the modern world (Mainguet 1999; Makeschin et al. 2001; Zhang 2002). In the developing countries, arable lands decrease as populations grow. Accordingly, a great amount of forests, grasslands, and wetlands are being exploited and intensively managed for cultivation, leading to the original vegetation

J. Zhang, *Coastal Saline Soil Rehabilitation and Utilization Based on Forestry Approaches in China*, DOI 10.1007/978-3-642-39915-2_4,

and soil structure being damaged and ultimately to soil salinization (Malcolm and Sumner 1998; Marcar et al. 1999; Zhang and Li 2002). It is estimated that, worldwide, saline land is increasing annually by 1–1.5 million ha (Kovda 1983).

In China, the situation is among the most critical. It is well known that China is the most populated country in the world, and with that population growth, there has been rapid socioeconomic development and urbanization, especially in the eastern coastal region. Sea-level rise has caused serious problems in coastal areas of the country, with seawater intrusion aggravating coastal areas of Liaoning Province, for example, and subsequent salinization has become extremely serious, especially in the areas of Jinzhou, Huludao, and Dalian. Salinization has also occurred in Hebei Province, where seawater intrusion has reached 32 km in Haigang District of Qinhuangdao City, and in parts of Funing County, covering an area of more than 300 km^2 and resulting in groundwater and land salinization in these areas.

Coastal erosion is even more severe in Shandong Province. Affected by changes of sea level, disastrous occurrences of seawater intrusion and ongoing erosion in the coastal areas of this province are worsening. The total area of seawater intrusion in the areas of Yantai, Qingdao, Weihai, and Rizhao has reached 649 km^2, and coastal erosion from Longkou City to Yantai City stretches for 30 km with a cumulative maximum erosion width of 57 m.

Meanwhile, the coastal city of Shenzhen is likely to suffer from major seawater intrusion in the near future. The city is currently enduring lowering levels of groundwater, which means it faces contamination with replacement seawater. There has been excessive use of groundwater in Shenzen since the 1980s, as tap water was not widely used until the same period as when the city witnessed urban expansion.

Furthermore, soil degradation is worsening due to environmental impacts and intensive management (Zhang 2008). Thus, it is vital to remedy and reclaim disturbed land, especially saline land, throughout the country. The key aspects for dealing with the issue are sound city planning, rational land-use systems, restoring plant communities, and improving soil structure.

4.1 Process of Urban Salinization

Urban salinity, like salinity in rural areas, is caused by rising water tables. In a balanced groundwater system, nearly all rainwater enters the soil and is largely used by vegetation; consequently, the water table usually remains many meters below the soil surface. Thus, a hydrological balance exists between the amount of water seeping into the soil and water that could drain away or be used by plants. With large-scale clearing of native vegetation in rural and urban areas, any water added to the system in excess of evaporation and plant needs may seep through the soil, adding to the groundwater and making the water table rise rapidly to the surface in low-lying areas. As the water table rises towards the land surface, it carries with

Fig. 4.1 Collect soil samples in Shanghai (Photograph taken by Zhang Jianfeng)

it salts that are dissolved from the soil and rocks. Evaporation of the water leaves these salts behind to accumulate in the soil at shallow depths, thus causing salinity problems (see Fig. 4.1).

There are two kinds of soil salinity: dryland salinity (occurring on land not subject to irrigation) and irrigated-land salinity. Both describe areas where soils contain high levels of salt. Usually, plants and soil organisms are killed, or their productivity is severely limited, on salt-affected lands (Marcar et al. 1999). Irrigated-land salinity results from the clearance of trees and replacement with shallow-rooted crops, overwatering, seepage from irrigation channels, and impaired natural drainage. The main cause of high water tables and dryland salinity is land-use change, particularly the broadscale replacement of deep-rooted native vegetation, such as trees and perennial native grasses, with shallow-rooted annual agricultural crops and pastures that use less water.

In recent years, it has been recognized that many other land management practices aside from irrigation and clearing for agriculture may contribute to rising water tables and saline seepage in many areas. Such practices include excessive watering of lawns and gardens; tree clearing for farming and/or urban development; irrigation of public recreation areas; poorly designed septic and sewerage systems; and the development of roads, bridges, railways, and flood control banks, all of which interfere with natural surface drainage patterns and induce ponding of surface water (Makeschin et al. 2001).

4.2 Impacts of Salinity on Plant Growth and Urban Structure

Soil structure in sodic soils is poor and permeability is low. If a sodic clay layer occurs near the surface of sodic soils, it often acts as a barrier to the roots. Hence, most roots are restricted to the topsoil above the claypan because the movement of water, nutrients, and gases in sodic B horizons is too slow. When dry, the B horizon can be so hard that it physically restricts root penetration. The overall effect on plant growth is one of stress, similar to that caused by extremely dry or saline conditions.

High soil salinity (high EC) causes dehydration of plant cells, which in turn results in reduced plant growth and often death. Dissolved salts cause dehydration of plant cells because the osmotic potential of soil water is decreased. Water will flow from the high osmotic potential (low salts in plant cells) to low osmotic potential (high salts in soils). Thus, plants simply cannot extract water from soils when soil solutions have lower osmotic potentials than plant cells. The effect on plants is similar to drought stress. Yields of many crops are reduced when the soil extract EC reaches 4 dS/m. Yields will decline proportionately as EC levels increase above 4 dS/m. Some crops, such as sugar beets and cotton, are tolerant to EC levels between 4 and 8 dS/m. The growth and yields of most crops are affected at EC values of >16 dS/m.

The impacts of urban salinization on urban structure can be summarized as follows:

1. Damage to houses, buildings, and other structures caused by the deterioration of brick, mortar, and concrete due to saline water crystallizing in brickwork.
2. Corrosion of metal buried in the ground or set in structural concrete may also occur.
3. Shifting or sinking of foundations may result in structural cracking, damage, or collapse. Damage to heritage buildings may be of particular concern, and land values may be adversely affected by salinity.
4. Salt damage to roads and highways, including the breakdown of concrete, bitumen, and asphalt with associated pot-holing, cracking, and crumbling of the road base.
5. Damage to underground pipes, cables, and other infrastructures due to the breakdown of unprotected metal, cement, and other materials.
6. Loss of amenity in recreational areas, such as gardens and sports fields, due to the appearance of bare, exposed patches where grass and other plants cannot grow.
7. Failure of septic tanks caused by high water tables. This often leads to other environmental and health problems.
8. Adverse effects on the environment of local flora and fauna.

In the context of the abovementioned problems and issues, taking measures to prevent soil salinization in urban areas is therefore vital and significant.

4.3 Strategies for Dealing with Saline Soils

Salinity is not always related to the impact of dissolved ions but may also include dissolved gases (under pressurized conditions), which, once the pressure is released, may give rise to other problems, such as cracking of basement clayey rocks, case-hardening, and sodification. Consequently, while urban salinity may be broadly linked with rising local and/or regional groundwater levels, for a meaningful quantification of its impacts on the natural and built environment, a much better understanding of the chemical composition of such groundwater is needed. Such information is also needed to identify and implement appropriate control measures.

It has recently been demonstrated that most of the steps commonly proposed or implemented for reducing groundwater recharge (tree planting etc.) neither reduce the inflow nor retard the salinization process sufficiently in urbanized areas in the short-to-medium timeframe (Zhang 2008).

There are different ways to ameliorate saline soils. Successful remediation of highly saline–sodic, clayey soil via leaching requires the maintenance of some degree of hydraulic conductivity (HC) throughout the soil profile. It is generally thought that the degree of swelling in salt-affected soils can be manipulated by controlling EC and SAR. Ideally, SAR, or the ratio of sodium to calcium and magnesium cations, should be reduced prior to reductions in EC to prevent loss of HC caused by swelling and dispersion of soil clays. A study was carried out to examine the efficacy of several treatment combinations intended to manipulate saline–sodic soils, chemically and physically, during the early stages of remediation. Treatments included calcium chloride, gypsum, and bagasse, alone and in various combinations. Additionally, subsurface drainage and sprinkler irrigation systems were installed over the entire site. Subsurface drains and collection sumps were installed to overcome the seasonally high water table (fall/winter), and drained waters collected in sumps were hauled to a nearby deep injection well for disposal. A sprinkler irrigation system was installed to supply water during the dry season (spring/summer) to take advantage of the seasonally low water table.

A high incidence of middle- and high-canopy shrubs and trees left or planted on farms serves as physical windbreaks, and control of the wind in this manner helps to reduce the rate of evapotranspiration from crops growing on the land (see Fig. 4.2). This is necessary particularly in those farm holdings located in drier areas. Indeed, this notion that with more trees there is less evaporation and less salt accumulation in topsoil has been tested and shown by Zhang (2004).

Extensive site characterization led to the identification of sodium chloride as the primary salt contaminant in soils associated with oil production activities. The deleterious effects of salt contamination, particularly NaCl, upon a soil's chemical and physical properties are well known. Toxicity to plants as well as the deterioration of soil physical properties is a common occurrence. This can be evident as reduced infiltration and aggregation ultimately enhance the potential for soil erosion. It should be noted that the degree to which sodicity affects any given soil is influenced by factors such as soil mineralogy, pH, texture, and organic content.

Fig. 4.2 Trees planted in Yantai, Shandong (Photograph taken by Xu Jingwei)

Hence, improved irrigation and drainage systems are required to control the rise of the water table and environmental salt accumulation. For example, land-water could be irrigated with saltwater and freshwater alternately, surface mulch could be used, or surface soils could be furrowed after rain. Crops and grasses possess different tolerance levels for salt in different water and salt environments. Therefore, selected crops or grasses must be cultivated on the basis of different types and salinities of soils. In addition, agronomic measures could be used, such as delaying the time of seeding.

In China, salt-affected land is a widespread problem and includes areas that are important land sources for agricultural production. Successful remediation of saline–sodic soils depends on the ability to leach salts to a depth below the rooting zone. Maintenance of soil physical and hydraulic properties is of paramount importance in achieving this goal. Failure to do so negates any possibility of in situ remediation by means of leaching. It has long been recognized that leaching of sodium salts from a soil can result in the destruction of soil structure by means of clay dispersion and swelling when the leaching solution drops below a critical electrolyte concentration. Additional work has shown that the ratio of sodium to calcium in the soil solution plays an important role in determining this critical electrolyte concentration. In essence, it is important to maintain a high EC until SAR can be reduced to an experimentally determined level to avoid dispersion and swelling.

References

Kovda VA (1983) Loss of productive land due to salinazation. Ambio XII(2):91–93

Mainguet M (1999) Aridity drought and human development. Springer, Berlin/Heidelberg

Makeschin F, Landgraf D, Zhong Z (2001) Nachhaltige Agroforstwitschaft auf salinen Koestenboeden der Provinz Zhejiang. Wissenschaftliche Zeitschrift der Technischen Universität Dresden 4:102–108

Malcolm E, Sumner RN (1998) Sodic soils-distribution, properties, management, and environmental consequences. Oxford University Press, New York

Marcar N, Ismail S, Hossain A (1999) Trees, shrubs and grasses for saltlands. Australian Centre for International Agricultural Research, Canberra

Zhang JF (2002) Research on saline soil amelioration and forestation techniques. J Northeast Forest Univ 30(6):124–129

Zhang JF (2004) Agroforestry and its application in amelioration of saline soils in eastern China coastal region. Forest Study China 6(2):27–33

Zhang JF (2008) Principles and technologies of saline soil ecological amelioration. China Forestry Press, Beijing

Zhang JF, Li XF (2002) Strategies to reclaim and ameliorate saline soil in the Yellow River Delta Region. In: Chang Chi, Brian D (eds) International conference on environmentally sustainable agriculture for dry areas for the 3rd millennium proceedings. Dobing Enterprises, Canada, pp 264–269

Part II
Principles and Practice of Afforestation in Saline Soils

Chapter 5
Salinity Tolerance and Tree Planting Techniques

Abstract With rapid world population growth, the area of arable land is decreasing, and thus saline soil has become an important natural resource. However, to date, reclamation of salt-affected land has been limited owing to adverse conditions. Forestation is one way to utilize saline soil and can improve the quality of the soil as well. The mechanisms of salinity tolerance form the basis for tree species selection when planting in salt-affected soil, with different species having evolved different methods for coping with salinity. These methods fall into the following categories: salt exclusion (e.g., *Elaeagnus angustifolia*), salt secretion (e.g., *Tamarix spp.*), salt dilution (e.g., *Hordeum vulgare*), and salt avoidance (e.g., *Rhizophora apiculata*). Many factors should be taken into consideration when choosing trees for selected sites. For example, nursing seedlings near the site is helpful to enable trees to adapt to the adverse conditions; site preparation is crucial for growing seedlings and planting trees; and density/spacing—dependent not only on the tree species but also on the forest type, site conditions, and socioeconomic considerations—must also be taken into account. Finally, decision-making in terms of when to plant requires care and caution; choosing a date that avoids soil salt damage and promotes seedling survival and growth is essential.

Keywords Forestation • Saline soil • Saline soil management and utilization • Site selection • Tree species

In China, there is a large area of salt-affected land, for example, in Lop Nur and Turfan (both around 5,000 km^2), almost the same, and in Qaidam Basin (20,000 km^2) (Gong 1986). Owing to the adverse conditions of such soils, crops are unable to grow, but planting trees by suitable means is possible. Forestation is not only one way for saline–alkali soil utilization but can also improve soil properties and maintain ecological balance (Bohnert et al. 1995). Thus, it is important to discuss the principles and techniques of forestation in salt-affected soil.

J. Zhang, *Coastal Saline Soil Rehabilitation and Utilization Based on Forestry Approaches in China*, DOI 10.1007/978-3-642-39915-2_5,

5.1 Salinity Tolerance of Trees

Salts are a critical limiting factor for tree survival and growth, and so understanding the mechanisms of salinity tolerance in trees is fundamental when it comes to planting trees in salt-affected soil.

Excess salinity in soil is harmful to trees. On the one hand, it influences the survival of roots and refrains or hinders the absorption of mineral elements and water; while on the other hand, salts can directly injure roots, damage plant tissues, and disturb normal physiological activities of trees (Lieth and Masoon 1993).

Through a long history of natural selection, some tree species have evolved to become salinity tolerant, to varying degrees and through using different methods. For example, some tree species use salt-exclusion methods, in which they refuse to take up salts (e.g., *Elaeagnus angustifolia*) (Ma 1997). Others use salt secretion, which involves secreting salt from glands after absorbing it from the soil (e.g., *Tamarix* spp.). Another method is salt dilution, which means the tree can dilute salt in order to maintain lower levels of it in cells (e.g., *Hordeum vulgare*). And lastly, some species use salt-avoidance techniques (e.g., *Rhizophora apiculata*). Certain tree species are characterized by their ability to use a variety of methods in a comprehensive approach to salinity tolerance (e.g., *Fraxinus* spp.) and in doing so can endure salt stress to become important pioneer tree species for planting in saline soils (Fan 1990).

Salinity tolerance in trees means they can survive and grow well to some extent after being planted for 1–3 years in saline–alkali soil, and their traits of salinity tolerance can be described by surveying survival rate and growth or productivity. Different plants have different capacities to tolerate salt. Accordingly, the same tree species may show different abilities when growing in salt-affected soil depending on age, salt type, soil characteristics, etc. When selecting trees for planting in saline soil, it is very important to understand their characteristics in terms of salinity tolerance (Staples and Toenniessen 1984).

5.2 Principles and Techniques of Afforestation in Saline Soils

The principles of forestation in saline–alkali soils involve (1) selecting suitable tree species for the intended site according to the rules of eco-economics; (2) preparing the site to ameliorate soil conditions; (3) nursing seedlings by means of effective propagation; (4) planting at the optimum moment in time to reduce salt damage; and (5) regular maintenance to help trees grow and improve the soil quality. Working to these principles helps to obtain ideal ecological conditions as well as satisfactory economic benefits (Zhang et al. 2010).

In order to ensure the success of forestation in saline soils, a necessary precondition is to choose suitable trees for the targeted patch of land (good species–site matching), especially given that trees grow slowly and the cost of planting is

Fig. 5.1 Trees of *Tamarix* growing in Hebei (Photograph taken by Zhang Jianfeng)

quite high. If conditionally unfit trees are planted, they will not grow well and may not even survive at all. Therefore, tree species selection must be carried out carefully, and the traits of the trees should also be taken into consideration (Zhang et al. 2003a). Techniques for forestation in saline soils vary with different climates, sites, and tree species, and these techniques are now described for the cases stated above.

5.2.1 Tree Species Selection

It has already been highlighted as to why tree species selection is important. The following three points provide more detail regarding which trees can be chosen depending on the specific conditions of the site.

1. *Strong salinity tolerance.* When soil salt concentration in a particular site is high, only trees with strong salinity tolerance should be planted. Usually, when the soil salt content is greater than 1 %, *Nitraria* spp. in coastal regions can be grown; when the salt concentration is about 0.5 %, *Tamarix* spp. (see Fig. 5.1), *Lycium chinense,* and *Amorpha fruticosa* could be adopted; and when it is below 0.4 %, *Fraxinus velutina*, *Ziziphus jujuba*, *Rhus chinensis*, *R. typhina*, *Ailanthus altissima*, *Melia azedarach*, *Sophora japonica*, *Ulmus pumila*, *Robinia pseudoacacia*, among others, could all be planted.

2. *Strong drought and waterlogging tolerance.* Salinization is always related with drought and waterlogging. In arid and semiarid regions, rainfall is low, and when surface water cannot meet the demands of plants, underground water is instead exploited. When this water is transported upwards, it carries with it minerals that are taken up by plants, and extra salts subsequently accumulate in surface soil layers. Hence, drought and salinization often occur at the same time. Therefore, when selecting tree/shrubs species, their capacity to deal with drought and waterlogging must also be taken into account (Zhang 1997).
3. *Fast growth and easy propagation.* Usually, plant communities in saline–alkali soils are rare, which is one of the causes of high evaporation and salinization. Thus, keeping the vegetation coverage as high as possible is necessary in salt-affected lands. Moreover, a higher coverage of vegetation can combat weeds and increase soil organic matter content, which is helpful for trees to continue growing (Jiang 1999).

Additionally, targeting particular tree species can help meet the demands of socioeconomic development, especially by utilizing local species. Besides the abovementioned biological factors, other considerations, such as religious taboos and planting habits, are also important when selecting tree species for forestation.

5.2.2 Improvement of Sites to Reduce Salt Content

The aim of forestation in saline soil is to utilize natural resources and increase land production. Within this process, there are a number of hindering factors, of which the most critical is too much soluble salt in the soil. After tree species have been selected, improving the site conditions is very important for promoting their subsequent growth. The main purpose of such improvements is to reduce soil salinity (Sharma and Gupta 1986), and there are a number of different categories of approaches, as outlined below.

The first is water conservation, which is often used to build sound irrigation systems in order that drought can be alleviated and waterlogging avoided. As discussed above, salinization is always related to conditions of drought and waterlogging, and in view of this, constructing complete water conservation facilities—with the principal aim being to control the underground water table—is necessary to combat salt accumulation at the soil surface.

Another category of measures is farming practices, including deep plowing, intensive site preparation, and the killing of weeds. Such approaches aim to improve soil physical properties and cut soil capillary action, thus preventing salt from rising and decreasing soil salinity. Finally, there are biological measures, which are currently most effectively used in practice and include the application of green manure to increase soil organic content and intercropping with plants of *Leguminosae* to enhance biological nitrogen fixation and soil fertility.

All of the above measures should be operated according to local site conditions, the tree species planted, the forest type, and economic benefits. When the plantation is intensively cultivated, more silvicultural and management measures could be implemented in order to promote faster tree growth.

5.2.3 *Seedling Nursing*

Practices related to the nursing of seedlings play an important role in determining the quality of forestation. Salinity tolerance in trees is obviously required when planting in saline soil; however, despite careful selection for this trait, some species still may not adapt well to the adverse conditions. The key to resolve this problem is nursing the seedlings near the woodland, where they are exposed to background levels of salt stress and can adapt to the conditions gradually (Zhang et al. 2003b).

Although clones have been used in many plantations throughout the world, most salinity-tolerant trees are still propagated by seeds (Zhang et al. 2003c). Thus, site selection and site preparation of the nursery are extremely important. The nursery needs to be located where salinity is lower, the underground water table is higher, the soil texture is better, and an irrigation system is available. In addition, the plot should be located near roads and villages so that it can be cared for intensively and the seedlings can be easily transported.

Site preparation of the nursery includes plowing, applying fertilizer (especially base fertilizer), additional manuring, weeding, as well as bed preparation. Regarding the latter, high nursery beds are suited to locations where rainfall is sufficient, but drainage is poor. Usually, the height of the bed will be around 10–25 cm above the land. Low beds are more suited to places where it is dry or salinity levels are particularly strong. The depth of beds tends to be around 15–25 cm lower than the land. At present, container nurseries are also becoming popular, which can help avoid excessive soil salt and make management easier.

5.2.4 *Planting*

One of the key steps in forestation is planting, and much like seedling nursing, site preparation before planting is extremely important. The specific measures themselves are almost the same, while in urban areas saline soils are sometimes removed or replaced by rich soil in order to ensure the survival and rapid growth of the trees.

Planting time is a critical factor when forestation is implemented in saline–alkali soil—usually to coincide with lower levels of soil salinity at the surface and thus providing the trees with a better chance to grow normally in the initial stages. Specifically, planting trees in spring as early as possible is the best approach because

Fig. 5.2 Planting trees in the Hangzhou Bay area (Photograph taken by Shan Qihua)

in spring salts have been leached into the deeper soil layers and roots will therefore not be damaged by surface soil salt after being planted. Indeed, most trees can be grown in spring (see Fig. 5.2).

For some trees, planting in summer (rainy season) is feasible, as salts are leached with water and temperatures are higher, which are suitable conditions for seed germination. Some species, such as *Nitraria* spp., *Tamarix* spp., and *Amorpha fruticosa,* can be grown by direct seeding. Furthermore, the practice of planting container seedlings can also be adopted in this season.

Density is a factor closely related with growth of the stand. A sensible density can promote growth and the successful formation of the forest environment. The particular spacing used depends on the species, as well as the purpose or intended product of planting, the planting method, site conditions, and so on. However, as a higher plant coverage can limit soil evaporation and thus help prevent salt returning to the soil surface, a high density of planting is generally recommended. Usually, a spacing of 1 × 1 m for *Nitraria* spp., *Tamarix* spp. (Zhang 2002), *Amorpha fruticosa,* and *Hippophae rhamnoides* is employed; for *Elaeagnus angustifolia, Pyrus betulaefolia,* and others, a spacing of 1 × 2 m is adopted; for *Pinus thunbergii* and Euphrates poplar, 2 × 2 m is suitable; and for trees like *Ulmus pumila, Sophora japonica, Robinia pseudoacacia, Rhus chinensis, R. typhina, Ailanthus altissima,* and *Fraxinus velutina,* a spacing of 2 × 3 m or 3 × 3 m is preferable.

Actually, a mixed forest is ideal when site conditions allow. Some types of mixed forest have been used to good effect, for example, *F. velutina* × *A. fruticosa*; *P. thunbergii* × *R. typhina*; and *S. japonica* × *R. pseudoacacia*. The results of the surveys have shown that trees grow faster and the state of stands is better in mixed forests than in monoculture stands (Xu 1996).

Based on these principles and techniques, a large area of saline land has been reclaimed, and a great number of plantations have been established in China, for example, the "Three North" shield belts, the Yellow River delta area mixed forests, and the windbreak forests of Xinjiang. Another significant achievement is that a coastal protection forest system, which is one of a number of national planting projects, is all but complete. These forests cannot only improve the environment but also deliver products, thus playing a significant role in sustainable regional socioeconomic development.

Reclamation of saline–alkali soil is becoming more and more urgent as the world's population grows and the availability of resources approaches crisis point. Planting trees is one way to utilize saline soil, and the basis of success in this regard is to take advantage of the mechanisms of salinity tolerance in trees, that is, tree species selection. Many factors should be taken into consideration when choosing trees for selected sites. Nursing seedlings near the site is helpful for trees to adapt to the adverse conditions. Site preparation is crucial for growing seedlings and planting trees. The chosen density of planting is dependent on the tree species being used, the forest type, site conditions, and economic factors. Decision-making in terms of planting time is important and must be considered carefully, although choosing a planting date to avoid soil salt damage and promote the growth and survival of seedlings is generally the best approach.

As science advances, new techniques can be applied to analyze the mechanisms of salinity tolerance and breed new salt-tolerant varieties (Zhang 1990). With the help of techniques such as genetic engineering and tissue culture, many new lines, clones, and varieties with stronger salt tolerance have been developed, which is helpful in terms of rapid forestation and the amelioration of saline soils.

References

Bohnert HJ, Nelson DE (1995) Adaptation to environmental stresses. Plant Cell 7:1099–1111

Fan BM (1990) Preliminary study on salinity tolerance of *Fraxinus velutina*. Dissertation, Shandong Agricultural University, Taian

Gong HZ (1986) Silviculture in saline soil. China Forestry Press, Beijing

Jiang ZH (1999) Forest ecological engineering projects and the sustainable development at Yellow River delta region. Forest Res 5:447–451

Lieth HA, Masoon AL (eds) (1993) Towards the rational use of high salinity tolerant plant. Kluwer, Amsterdam, Netherlands

Ma HC (1997) Preliminary study on salinity tolerant physiology of Euphrates poplar. Dissertation, Beijing Forestry University, Beijing

Sharma SK, Gupta IC (1986) Saline environment and plant growth. Kluwer, Amsterdam, Netherland

Staples RC, Toenniessen GH (1984) Salinity tolerance in plants strategies for crop improvement. Kluwer, Amsterdam, Netherland

Xu XG (1996) Assessment and early warning study at ecosystem of Yellow River delta region. Acta Ecologica Sinica 5:461–468

Zhang JF (1990) Study on plantlets cultivation of new variety 74009 of white elm. Dissertation, Shandong Agricultural University

Zhang JF (1997) Review of saline soil utilization. J Shandong Forest Sci Technol 4:22–25
Zhang JF (2002) Research on saline soil amelioration and forestation techniques. J Northeast Forest Univ 30(6):124–129
Zhang JF, Li JY, Xing SJ (2003a) Advances of research on the mechanism of plant salinity tolerance and breeding of salt-tolerant plants. World Forest Res 16(2):16–22
Zhang JF, Li JY, Xing SJ (2003b) Experiment of seed germination under salinity stress. J Northeast Forest Univ 4:79–80
Zhang JF, Li JY, Xing SJ (2003c) Experiment on seed germination of *Chionanthus retusa* and *Cedrela sinensis* under salinity stress. J Beijing Forest Univ 4:88–90
Zhang JF, Chen GC, Xing SJ et al (2010) Carbon sequestration of black locust forests in the Yellow River delta region, China. Int J Sustain Dev World Ecol 6:475–480

Chapter 6
Agroforestry Systems for Combating Coastal Saline Soil

Abstract Coastal regions play an important role in China. However, a number of environmental problems, particularly soil salinity, hinder the sustainable development of these regions. Salinity mainly results from tidal reach, seawater flooding, and seawater intrusion. Soil salinization is caused by water evaporation, along with a rise of the water table. Furthermore, overexploitation of groundwater, largely resulting from harmful farming systems, leads to seawater intrusion and an increase in salt concentration. Agroforestry systems consist of multiple ecosystems that can enrich soil fertility and prevent soil salinization, and thus, these systems help maintain biodiversity and enhance soil productivity. In a multiple ecosystem, different plants occupy various ecological niches within an area, both in space and soil depth. This setup allows competitive occupation in the same and in similar areas while reducing soil degradation through excessive extraction of water and nutrients from the same soil horizon. The most critical issues are selecting optimum tree species and rationally placing these species. High crop diversity also limits total loss of crops or yields resulting from diseases and pests. Several plant relationships offer opportunities for diversified and improved land productivity. Shelterbelts reduce wind rates and, consequently, lessen soil evaporation and erosion, enhancing pasture growth by up to 60 % on exposed sites and increasing crop yields by up to 25 %. Apart from intercropping with jujube, coastal regions can best be used as land resources through other agroforestry multiple ecosystems, such as forestry plus agriculture, forestry plus agriculture plus fishery, and forestry plus animal husbandry.

Keywords Agroforestry • Amelioration • Saline soil

Rapid population growth, a decrease in available arable land, industrial development, and environmental deterioration, particularly under the promotion of economic reforms and the Open Door Policy, have combined to push small farming systems in China to be on the verge of collapse.

J. Zhang, *Coastal Saline Soil Rehabilitation and Utilization Based on Forestry Approaches in China*, DOI 10.1007/978-3-642-39915-2_6,

Modern agroforestry is based on the principle of biological production, from which economic growth can be achieved through the processing industry and market system (Buck 1995). By combining production and market procedures, the advantages of natural resources, for example, land, climate, and water, along with social resources, such as labor, science, and technology, are maximized, while maintaining sustained economic, social, and ecological benefits of agroforestry (Schultz 1995).

Based on this conceptual framework, various kinds of modern agroforestry systems have been designed and implemented, including home gardens that grow flowers, medicinal plants, and highly valued fruits and vegetables for the market; fish raised in rice fields; animals grazed on rangeland with dairy and milk processing; intercropping in plantations; and timber mill operations. These systems involve multiple production factors and components, such as cultivation, processing, and marketing of trees, crops, domesticated animals, and aquatic species. The operations of these systems have been carefully arranged and conducted according to the biological characteristics and economic values of their components; the intermediate reactions among these systems have also been considered (Zhang 1997a). The components concerned have been purposefully integrated into production systems in appropriate time sequences, spatial arrangements, and numerical proportions, thus allowing them to fully perform their biological production role and create favorable conditions for successive components. Correspondingly, processing and marketing systems can be established according to the layout and quantity of biological products (yields). Consequently, an entire rural economy combining agriculture, industry, and trade can be formed and maintained. This setup increases output per unit of land, augments general biomass, creates employment opportunities, raises total income, improves living standards, and promotes rural development in these regions.

6.1 The Significance of Developing Agroforestry

China is the most populated country in the world; the population of mainland China in 2012 was 1.3 billion. Its land area is approximately 9.6 million km^2, and the proportion of arable land is low, covering approximately 10 % of the total land area. On average, approximately 0.09 ha of tilled land per capita can be observed. Furthermore, arable land is unevenly distributed across the country. Forests cover 13.92 % of the entire land area, corresponding to 134 million ha or 0.095 ha per capita (Xu 1995). Hence, China is a country with one of the lowest coverage of forest on Earth. As a result of population growth, Chinese people have for a long time been facing the challenges of a heightened demand for wood and increased food consumption (Zhang 2002). Grain production is the most important factor in terms of providing food for the growing population; however, the available arable land generally has poor productivity. Unfavorable climate and natural hazards, such as floods, drought, sandstorms, soil erosion and degradation, salinization,

Fig. 6.1 Poplar trees as windbreakers and shelterbelts in Shandong (Photograph taken by Zhang Jianfeng)

and desertification, along with other difficult-to-define conditions, are historically attributed to deforestation (Wang 1996). These major factors have caused the low productivity of land resources.

Land resources continue to be improperly utilized for industrial and other nonagricultural uses, such as for urban expansion and the construction of roads, railways, and airports. According to the Agricultural Natural Resources and Planning Institute of the Chinese Academy of Agricultural Science, the annual average net loss of cultivated land is estimated to be approximately 0.2 million ha per year. Consequently, the conflict between population growth and land reduction is becoming increasingly serious. The significance of land use and other related issues has certainly become more apparent (Zhang 2003).

The aforementioned situation and the tendency towards socioeconomic development emphasize the need to apply and develop agroforestry in China (Zhang 1994). Agroforestry, as an integrated land-use system, is an obvious means of fulfilling multiple objectives. Agroforestry has numerous advantages and implications (Buck 1995). The value of trees on farms can far exceed the value offered by woodlots and plantations alone. Trees act as windbreaks and shelterbelts, delineate boundaries, and provide shade, ornamentation, and seclusion around homesteads (see Fig. 6.1). Apart from providing poles, stakes, timber, and fuel, trees also supply cash crops, fodder, fruits and nuts, dyes, gums, resins, fiber, and medicines, depending on the localities and species (Dabbert 1995). Fodder and food trees can help provide balanced meals during dry seasons, when other kinds of food are scarce. Compared with arable and pastoral crops, trees have deeper root systems that can exploit moisture and nutrients at a greater depth in the soil; therefore, humans should take advantage of these available resources. Furthermore, tree roots reduce the leaching of nutrients from soil after heavy rains and conserve soil moisture by adding mulch and litter. In addition, leguminous trees can fix nitrogen and enrich soil fertility.

Agroforestry not only offers a suitable balance of production between wood and food but also increases the total productive capacity per unit area of land (Dabbert 1995). Considering the environmental costs of flooding and erosion, the economic merits of agroforestry become more apparent and attractive.

6.2 Eastern China: The Main Environmental Problems

The provinces of Liaoning, Shandong, Jiangsu, Shanghai, Zhejiang, and Fujian represent the coastal regions in eastern China. Given the geographical location and climatic conditions of these regions, their economy is relatively developed compared with other areas in China. However, their population density is relatively high; for example, in Zhejiang Province, the population density is 480 people per km^2, compared with the average of 74 people per km^2 in the entire country. With population growth and socioeconomic development, environmental degradation problems, such as pollution, noise, salinization, and drought, are becoming more serious (Zhang 2001). Of these issues, salinization is a critical problem mainly related to the increasing population and limited land resources in the region (Makeschin et al. 2001; Zhang and Li 2002).

Agroforestry is an important strategy for coping with salinization (Zhong 1998). The benefits of agroforestry practices are significant from the economic and environmental perspectives. Agroforestry can increase farm profitability in two main ways: (1) The total output per unit area of tree/crop/livestock combinations is more than that of any single component alone (Marcar et al. 1999). (2) Crops and livestock, which are protected from the damaging effects of wind, are more productive (Zhang 1997b). Agroforestry helps conserve and protect natural resources. For example, it controls soil erosion and creates habitats for wildlife (Qureshi and Barrett-Lennard 1998). The benefits of agroforestry lead to a substantial improvement in the economic and resource sustainability of agriculture (Zhong 1996, 1998).

6.3 Salinity Sources and the Salinization Process in Eastern China

The saline soil discussed in this chapter includes coastal solonchak, beach solonchak, and coastal salinized meadow. Salinity in coastal soil results mainly from the following phenomena: (1) Seawater directly reaches land via the tides, and the salt permeates into the ground. (2) Salt moves from the sea into soil via wind transport, particularly during typhoons, when water seeps into the ground. An example is given in Table 6.1 to describe this phenomenon (Li and Xing 1994). (3) Seawater intrusion,

Table 6.1 NaCl distribution with distance to sea

Measurement spot	Distance to sea (m)	NaCl content in water (g/mL)	NaCl content in pine needles (mg/g)
1	100	0.182	5.76
2	300	0.097	3.17
3	500	0.074	1.69
4	700	0.038	1.71
5	900	0.039	0.42

which is caused by the overexploitation of groundwater, as well as water table failure, disturbs the balance between freshwater and seawater (Qureshi and Barrett-Lennard 1998). Seawater intrusion has recently been discovered in Shandong Province and mainly occurs in the Jiaozhou Gulf area covering the counties of Laizhou, Jiaonan, Jiaozhou, and Shouguang.

It is clear that salinity and the dynamics of salinization are relatively different along coastal lines and in various locations (Landgraf et al. 2002; Makeschin et al. 2001), particularly in sites where soil properties vary widely because of specific topography, parent materials, and farming systems, as shown in Tables 6.2 (Project of Seabeach Resources Investigation 1988) and 6.3 (Li and Xing 1994).

Salinization has various causes (Zhang 1997a). In beach areas, salt comes from seawater. In fields far from sea beach land that have been cultivated for a long time, salt results from irregular precipitation throughout the year and a high groundwater table, particularly in spring, when the weather is mostly characterized by dryness and winds. During this period, increased evaporation is observed. This phenomenon results from the rise of groundwater containing soluble salts, which in turn accumulate in subsoil after the water has gone. The loss of vegetative cover enhances evaporation and promotes salinization (Qureshi and Barrett-Lennard 1998). Salinization degrades soil properties and fertility, such that soil organic matter is decreased and soil is subjected to damage from intense tropical rainfall. With few organic substances in the soil, raindrops move clay particles and clog soil pores, thus inhibiting water infiltration and increasing runoff and erosion (Makeschin et al. 2002). Finer-grained eroded sediments damage aquatic productivity and bury what were once freshwater and near-shore marine aquatic breeding grounds. The remaining coarser sandy materials from weathered granite yield soil with low fertility. The most direct effect of this disastrous loss of productive soil is reduced agricultural production (Doran and Turnbull 1997). Farmers growing crops under such conditions face tremendous difficulties, even with the use of commercial fertilizers. Soil from weathered granite is poor at holding plant nutrients found in commercial fertilizers. In addition, fertilizers are easily washed off the land during heavy rains.

Stripped of vegetation that would otherwise have absorbed or slowed down the flow of water, water pours rapidly into streams and rivers, cutting deep ravines in the soft, deeply weathered granite. Given the aforementioned scenarios, establishing a sound vegetation cover system is necessary in this region (Zhang and Li 2002).

Table 6.2 Salinity, nutrients, and other physical properties of beach solonchak in Zhejiang Province

Soil	Soil particle >0.01 mm (%)	Soil particles <0.01 mm (%)	Texture	Salinity (0–100 cm) (%)	Organic matter (0–20 cm) (%)	N (0–20 cm) (%)	P_2O_5 (0–20 cm) (%)	K_2O (0–20 cm) (%)	C/N
Sandy beach	87.01	12.99	Sandy	0.671	0.62	0.034	0.12	1.89	10.58
Muddy beach	44.23	55.77	Loam	1.365	1.26	0.069	0.13	2.50	10.59
Clay beach	14.56	85.44	Clay	1.763	1.60	0.091	0.13	2.90	10.20

Table 6.3 Soil salinity distribution in the coastal region of Shandong

Investigation area	Soil salinity (0–20 cm) (%)	Soil salinity (20–40 cm) (%)	Soil salinity (40–60 cm) (%)	Soil salinity (60–80 cm) (%)	Mean soil salinity (%)
Hekou	2.15	2.58	3.10	3.21	2.76
Zhanhua	2.53	3.02	3.22	3.56	3.11
Dongying	0.81	1.60	1.86	2.10	1.59
Shouguang	2.60	2.90	3.12	3.34	2.99
Wudi	2.58	2.58	3.20	3.68	2.25

6.4 Effects of Agroforestry on Ameliorating Saline Soil

Even though the definition of agroforestry varies among authors, it can be simply defined as the use of trees on farms. The International Center for Research in Agroforestry (now known as the World Agroforestry Center) defines agroforestry as a dynamic and ecologically based natural resource management system that diversifies and sustains production by integrating trees on farms and in the agricultural landscape for increased social, economic, and environmental benefits for land users at all levels (Doran and Turnbull 1997).

Although plants compete with adjacent ones to a certain degree for vital resources, they can also help each other, particularly legumes, the residues of which are returned to the soil (Fig. 6.2).

Many trees can grow harmoniously with crops in the same field, and at the same time, a number of which can fix nitrogen, which in turn enriches soil when their residues decompose. Several trees have a light, thin canopy that allows adequate sunlight to filter through them to the crops below (Prinsley and Swift 1986). Crops that grow with trees save their moisture because the protection provided by the trees cuts down their evapotranspiration rate. Perennial indigenous agricultural crops, shrubs, and even timber trees on fields have been observed to drop a considerable amount of litter on the ground, which helps add humus and other organic inputs to the soil. This process also helps maintain a high level of fertility, and thus, soil is ameliorated as farming continues without any fallow period (Qureshi and Barrett-Lennard 1998).

These benefits also apply to subsequent nonleguminous crops that do not have such capability. Trees also improve soil in other ways. Leaf litter decomposes and adds tilth and nutrients. Even the root system, with its rhizobia and related communities of organisms, releases nutrients and improves soil structure upon decomposition, as shown in Tables 6.4 and 6.5 (Xing 2001). A number of trees "capture" nutrients that are too deep in the soil for crops to reach and bring these nutrients to the surface and back to the soil as litter, where crops can use them when litter decomposes.

Tables 6.4 and 6.5 indicate that *Nitraria* can effectively prevent salt accumulation in soil, particularly on the soil surface, which has remarkable physical properties. Soil nutrition in woodland is relatively better compared with that in bare land.

Fig. 6.2 Shelterbelts of poplar and *Amorpha fruticosa* in Shandong (Photograph taken by Zhang Jianfeng)

Table 6.4 Soil nutrition and salt content in different sites

Site	Soil depth (cm)	Organic matter (%)	Total N (%)	Available N ($\times 10^{-6}$)	P ($\times 10^{-6}$)	K ($\times 10^{-6}$)	pH	Salt content (%)
Woodland of *Nitraria*	0–0	2.66	0.05427	17.85	37.4	213	7.5	0.99
	20–40	2.20	0.04570	7.00	52.2	204	7.6	0.91
	40–60	0.99	0.02285	4.20	3.2	478	7.9	0.68
Bare land	0–20	0.45	0.03327	7.30	2.6	273	8.2	1.07
	20–40	0.78	0.02856	4.10	0.9	186	7.7	0.91
	40–60	0.61	0.02351	3.00	0.7	365	7.8	0.89

Table 6.5 Investigation of soil physical properties

Site	Soil depth (cm)	Soil moisture (%)	Bulk density ($g \cdot cm^{-3}$)	Soil porosity (%)
Woodland of *Nitraria* 1	0–0	24.92	1.17	55.85
	20–40	23.38	1.25	52.83
	40–60	24.56	1.35	49.06
Woodland of *Nitraria* 2	0–20	23.09	1.27	52.08
	20–40	26.99	1.28	51.70
	40–60	26.74	1.33	49.81
Bare land	0–20	27.30	1.46	44.91
	20–40	29.57	1.38	47.92
	40–60	29.59	1.44	45.66

Therefore, *Nitraria* is suitable and effective in improving soil quality and enriching soil fertility, which increases land productivity.

The perennial nature of indigenous herbaceous plants and trees provides continuous foliage and crown cover that prevent exposure of soil to direct radiation from the sun. This characteristic reduces evapotranspiration and surface soil caking, which consequently reduces water retention and degrades soil texture necessary for good growth and soil amelioration, respectively.

The high incidence of middle- and high-canopy shrubs and trees, left or planted on farms, serves as both physical and physiological windbreakers. Furthermore, taming the wind reduces the rate of evapotranspiration from growing crops, which is necessary, particularly in drier farm holdings in dry areas. More trees correspond to less evaporation and less salt accumulation in topsoil.

The previous chapter reported that *Tamarix* forest can effectively reduce soil salt concentrations, particularly in 0–40-cm soil organic substances. Soil moisture increases obviously, and soil capacity and total porosity are enhanced.

6.5 Agroforestry Systems in Eastern China

The expansion of farming into fragile land environments poses special problems for cultivating crops, including conservation and efficient use of limited water, maintenance of soil structure and fertility, wind control, and water erosion. One of the solutions to coping successfully with limited soil moisture is maintaining adequate soil organic matter and nitrogen (Landgraf et al. 2002). However, conventional methods of green manuring and mulch covering have generally been limited because of the difficulty in growing sufficient herbaceous biomass to return nitrogen and other nutrients to the soil and because of the competition among food crops for soil moisture, land use, and labor force. Thus, an optimum farming system involves practicing integrated land-use-type agroforestry systems.

6.5.1 Selecting Tree Species

Multipurpose trees, particularly nitrogen-fixing and salt-tolerant species that are well adapted to saline land conditions, provide several advantages compared with herbaceous sources of organic matter, nitrogen, and fodder. They have a higher feed value during the dry season. In addition, these trees can be grown at interstitial locations on the farm or with the crops without replacing them. If appropriately selected and properly placed with respect to specific crops, multipurpose trees can offer microclimatic benefits, such as wind-breaking effects and reduced evaporation (Zhang 1997a). Moreover, these trees can provide food, fuelwood, building materials, fodder, and other directly useful by-products while performing their ecological

Table 6.6 Tree species and their different degrees of salinity tolerance

Soil salinity (%)	Tree species
>0.6	*Nitraria* spp., *Casuarina equisetifolia*
0.4–0.6	*Tamarix spp., Lycium chinense, Euphrates poplar*
0.3–0.4	*Ulmus pumila, Amorpha fruticosa, Hippophae rhamnoides, Robinia pseudoacacia, Rhus chinensis, R. typhina, Fraxinus velutina, Eucalyptus botryoides*
0.2–0.3	*Sophora japonica, Ailanthus altissima, Ziziphus jujuba, Celtis tetrandra, Pistacia chinensis, Salix* spp., *Morus alba*
<0.2	*Pinus thunbergii, Populus* spp., *Melia azedarach, Sapium sebiferum, Albizia julibrissin, Taxodium ascendens, Elaeagnus angustifolia, Pyrus betulaefolia, Cyclobalanopsis glauca, Platycarya strobilacea*

roles on the farm. Lastly, as a form of standing capital, multipurpose trees can serve as sources of convertible savings for emergency needs, offering potential and attractive incentives to individual farmers who plant trees.

Trees are permanently grown; thus, the period of rotation is relatively long. Selecting suitable tree species is necessary in addressing adverse soil conditions (Marcar et al. 1999). The principles of species selection are as follows: (1) stronger tolerance to salinity to survive under salt stress, (2) rapid growth to cover the land and reduce evaporation as early as possible, and (3) multipurpose to allow farmers to obtain economic benefits apart from ecological benefits.

To assist with the soil salinity being such a major hindrance to establishing plantations, trees with different degrees of salinity tolerance are listed in Table 6.6 (Zhong 1996, 1998).

6.5.2 *Shelterbelts in or Around Farmlands*

Planting trees in or around fields to act as wind barriers, which is one of the main agroforestry systems, has been practiced in China for a long time (Fig. 6.3).

Shelterbelts can reduce the wind rate, and consequently lessen soil evaporation and erosion, enhance pasture growth by up to 60 % on exposed sites, and increase crop yields by up to 25 % (Doran and Turnbull 1997). Choosing the most suitable plant species for the situation is key, and some examples are provided below. However, note that some of the species mentioned do not grow well in all regions.

1. *Eucalyptus botryoides*. Suited for coastal sites and coppices; produces ground durable timber.
2. *Robinia pseudoacacia*. Suited for saline soil and coppices; fixes nitrogen; helps enrich soil fertility; prevents soil erosion. This tree can also be trimmed for stock fodder, regardless of whether the leaves are fresh or dry.
3. *Populus euramericana*. Has narrow crown varieties and medium density wood, which can be trimmed for stock fodder. This tree is tolerant to droughts.

Fig. 6.3 Shelterbelts of *Fraxinus velutina* on farmland (Photograph taken by Zhang Jianfeng)

4. *Salix matsudana.* Very fast growth rate; easily established from cuttings; can be trimmed for stock fodder. However, this tree possesses moderate susceptibility to *Nematus* spp. sawfly.
5. *Fraxinus velutina.* Grows fast and has high tolerance to salinity and drought.
6. *Casuarina glauca.* Tolerates moderate peaty or saline soil and requires trimming.

If established sensibly, shelterbelts can reduce wind speed by 50 % on the leeward side over a distance equivalent to eight times the height of the shelterbelt on a flat terrain (Qureshi and Barrett-Lennard 1998). Shelterbelts should be placed at intervals of six to ten times the height of each belt for maximum effect. Planting at 4 m and 2–3 m intervals is suitable for farmland shelterbelts and horticulture shelterbelts, respectively. Shelterbelts should normally be at 90° to the prevailing wind. Woodlots on suitable sites are more productive. Bare-stemmed species, such as *Eucalyptus*, can be supplemented with a second row of shrub species.

Planting trees that can form shelterbelt nets along roadsides, streams, or fence lines exhibits positive effects. The usual distance between main shelterbelts is 100–200 m, whereas the distance between secondary shelterbelts is 200–300 m.

6.5.3 Intercropping Agricultural Crops with Ziziphus jujuba

Ziziphus jujuba is popularly grown in China as a traditional fruit tree species. This species thrives from temperate to subtropical regions and is an ideal intercropping tree species, with a distribution center in Shandong and Liaoning. In practice,

Fig. 6.4 *Ziziphus jujuba* intercropping in the Yellow River delta region (Photograph taken by Zhang Jianfeng)

this tree can grow in unfavorable environments, such as in areas with annual precipitation of 400–500 mm and in soil with a certain degree of salinity (Zhang 2001). Moreover, jujube does not deploy leaves until the first 10 days of June and also has a thin crown. Intercropping space between two sprouts can range from 1.5 × 1.5 m to 3 × 3 m. Jujube enters the fruiting stage when it is about 4 or 5 years old, and the average yield can reach up to 6,000 kg/ha (Zhang and Li 2002). A new variety of jujube, called winter jujube, is being cultivated in Shandong. The fruits of this variety can be harvested from September to November and have much higher vitamin C and sugar content (Fig. 6.4).

In an intercropping system, the type of mixed planting of dates with wheat, soybeans, or peanuts results in increased yield for both crops. The yield of intercropped wheat can reach up to 3.8 t/ha, which is higher than that from a pure wheat field. Consequently, the economic benefit increases by more than 100 %. Intercropping with jujube has now exceeded 7,000 ha of land area.

Apart from intercropping with jujube, other agroforestry multiple ecosystems, such as forestry plus agriculture, forestry plus agriculture plus fishery, and forestry plus animal husbandry, have also demonstrated model functions, which are the most appropriate means of using land resources in this region. Intercropping crops on forest glades; pisciculture in rivers, ponds, and penstocks; and growing hydrophilic crops such as fodder and lotus rhizome all make up an agroforestry system wherein one component holds an association with another (Zhang 1997c). Consequently, the ecosystem is improved, soil salinity becomes the priority, and the derived economic benefit increases by as much as fivefold in some sites compared with monocrop farming.

The coastal regions in eastern China play an important role for the entire country. However, a number of environmental problems, particularly soil salinity, hinder their sustainable regional development. Salinity mainly results from tidal reach, seawater flooding, and seawater intrusion. Soil salinization is caused by water evaporation and the rise of the water table. Harmful farming systems result in overexploitation of groundwater, which leads to seawater intrusion and increased salt concentrations. Agroforestry systems consist of multiple ecosystems that can enrich soil fertility and prevent soil salinization, thus helping maintain biodiversity and enhancing soil productivity.

In multiple ecosystems, different plants occupy various ecological niches within an area, both in space and soil depth. This setup allows competitive occupation in the same and in similar areas while reducing soil degradation through excessive extraction of water and nutrients from the same soil horizon. Selecting optimum tree species and placing them sensibly are the most critical issues. High crop diversity also limits total loss of crops or yields caused by diseases and pests.

Several plant relationships offer opportunities for diversified and improved land productivity. Shelterbelts can reduce wind rate, and consequently lessen soil evaporation and erosion, enhance pasture growth by up to 60 % on exposed sites, and increase crop yields by up to 25 %. Apart from intercropping with jujube, other agroforestry multiple ecosystems, such as forestry plus agriculture, forestry plus agriculture plus fishery, and forestry plus animal husbandry, offer the most suitable means of using the land resources in eastern China.

The presence of a high diversity of complex plant relationships and interactions attracts other species of plants, insects, birds, and other animals, thus forming a complex farm ecosystem. This characteristic is one of the main benefits of these systems when they are managed effectively. By growing trees and crops harmoniously and by returning to the ground most of the nutrients taken from it, either by organic or inorganic means, then the system can be biologically sustained. The presence of several plant species makes agroforestry a biodiverse system. However, socioeconomic aspects will determine actual sustainability.

When forest products, such as timber for buildings, poles, fuelwood, fruits, nuts, oil, resins, and extracts for medicines, among others, are sold, they will help complement crop and animal production. Therefore, farmers can earn additional income from time to time. This income provides an incentive for encouraging farmers to plant trees on farmland.

References

Buck LE (1995) Agroforestry policy issues and research directions in the US and less developed countries: Insight and challenges from recent experience. Agroforest Syst 30:57–73

Dabbert S (1995) Agroforestry and land-use change in industrialized nations: a case study from northeastern Germany. Agroforest Syst 31:157–168

Doran JC, Turnbull JW (1997) Australian trees and shrubs: species for land rehabilitation and farm planting in the tropics. Australian Centre for International Agricultural Research, Canberra

Landgraf D, Lorei A, Makeschin F et al (2002) Carbon and nitrogen dynamics of dyked young saline soils in Zhejiang Province, P. R. China. In: Makeschin F (ed) Mitteilgn Dtsch. Bodenkundl. Gesellsch/Springer, Berlin

Li BH, Xing SJ (1994) Plants for planting in coastal land. Shandong Science and Technology Press, Jinan

Makeschin F, Landgraf D, Zhong Z (2001) Nachhaltige Agroforstwitschaft auf salinen Koestenboeden der Provinz Zhejiang (VR China). Wissenschaftliche Zeitschrift der Technischen Universität Dresden 50(4):102–108

Makeschin F, Fischer R, Zhong Z et al (2002) Bodenchememische Eigenschaften saliner Koestenstandorten der Provinz Zhejiang (China). In: Makeschin F (ed) Mitteilgn Dtsch. Bodenkundl. Gesellsch/Springer, Berlin

Marcar N, Ismail S, Hossain A (1999) Trees, shrubs and grasses for saltlands. Australian Centre for International Agricultural Research, Canberra

Prinsley RT, Swift MJ (1986) Amelioration of soil by trees-a review of current concepts and practices. Marlborough House, London

Project of seabeach resources investigation (1988) Report of seashore and seabeach resources investigation in Zhejiang Province. Ocean press, Beijing

Qureshi RH, Barrett-Lennard EG (1998) Saline agriculture for irrigated land in Pakistan: a handbook. Australian Centre for International Agricultural Research, Canberra

Schultz RC (1995) Agroforestry opportunities for the United States of America. Agroforest Syst 31:117–132

Wang B (1996) China's present forest ecological environment monitoring and environmental quality. World Forest Res 5:52–60

Xing SJ (2001) Study on salinity tolerance and planting techniques of *Nitraria* in worst salt affected soil. In: Liu MY (ed) The proceedings of international symposium on halophyte utilization and regional sustainable development of agriculture. Meteorology press, Beijing

Xu XQ (1995) Sustained development strategy on world forest environment. World Forest Res 2:1–7

Zhang JF (1994) Concept and properties of high yield, good quality, high profit forest. Shandong For Sci Technol (Special Collection) 24:30–34

Zhang JF (1997a) Study on optimum model of ecoagriculture in northern Shandong. Res Ecoagriculture 4:26–29

Zhang JF (1997b) Economic profit analysis of agroforestry systems in Huang-Huai-Hai plain. J Ecol Econ 4:12–14

Zhang JF (1997c) Review of saline soil utilization. J Shandong For Sci Technol 4:23–25

Zhang JF (2001) Principles and practice of afforestation in saline soil in China. In: Liu MY (ed) The proceedings of international symposium on halophyte utilization and regional sustainable development of agriculture. Meteorology press, Beijing

Zhang JF (2002) Research on saline soil amelioration and forestation techniques. J Northeast ForUniv 30(6):124–129

Zhang JF (2003) Advances of research on the mechanism of plant salinity tolerance and breeding of salt-tolerant plants. World ForRes 16(2):16–22

Zhang JF, Li XF (2002) Strategies of saline soil reclamation and amelioration in Yellow River Delta Region. In: Li DJ (ed) The proceedings of international symposium on agriculture for food and energy. Science press, Beijing

Zhong ZK (1996) Study on agroforestry and its effects in sands in Hangzhou Gulf. In: Agricultural comprehensive reclamation techniques in newly formed seabeach in Zhejiang. China Agricultural Science and Technology Press, Beijing

Zhong ZK (1998) Agroforestry in coastal saline soils in Zhejiang, China. Zhejiang Academy of Forestry, Hangzhou

Chapter 7
Strategies for Reclaiming and Ameliorating Saline Soil in the Yellow River Delta Region

Abstract The Yellow River delta, located from 37°21′ to 38°12′N and 118°07′ to 119°10′E, is one of the largest undeveloped deltas (a total land area of 12,000 km^2) in the world. With the increase in population and reduction of arable land, this huge region, which includes seven counties and 153,000 ha of national wetland reserve, has recently drawn the attention of both the government and other organizations. The annual increment of newly formed land (alluvium) is 2,000 ha. Salt-affected soil covers an area of 4,430 km^2. Saline soil has a high concentration of soluble salt that reduces the ability of plants to absorb water and nutrients. Based on specific climatic and environmental conditions, strategies for reclaiming and ameliorating saline soil in this region are discussed in this chapter. In the recharge area, improved drainage and water-efficient crop management practices could reduce the amount of water entering the groundwater system. In the discharge area where salinity is observed, establishing plant growth is important. Suitable cultivation systems, including leaching, tillage, and drainage, are practical strategies for ameliorating salt-affected soil. For several specific situations, applying amendments that improve soil structure and properties is effective. Moreover, adding manure and other organic or chemical fertilizers can increase production by enhancing soil fertilization. Planting suitable crops, including corn, grass, and forage, can directly produce harvests in a short period, without increasing land coverage. Reforestation is one of the vital measures for coping with environmental degradation and promoting regional sustainable development. The species involved could include Chinese ash, tree of heaven, Chinese sumac, Chinaberry tree, Siberian elm, and black locust.

Keywords Saline soil • Land reclamation • Soil amelioration • Yellow River delta

The population in Shandong Province, East China, is 93 million, but its total land area is only 153,000 km^2. As a result of socioeconomic development, use of arable land has been converted from agricultural to other purposes. At the

J. Zhang, *Coastal Saline Soil Rehabilitation and Utilization Based on Forestry Approaches in China*, DOI 10.1007/978-3-642-39915-2_7,

same time, environmental problems are becoming increasingly serious because of irregular natural resource exploitation and utilization. Consequently, a number of land resources are being degraded and salinization has become a crucial issue in several regions, particularly in dry areas. A number of scientists have conducted research on this topic (Pavelic et al. 1997; Webb and Price 1994). This chapter discusses methods for reclaiming and ameliorating saline soil in the Yellow River delta (YRD) region in Shandong Province.

7.1 Location and Natural Conditions of the YRD

The YRD in Shandong Province (37°21′–38°12′N, 118°07′–119°10′E) is one of the largest deltas in China, covering an area of 12,000 km^2. The region includes seven counties and 153,000 ha of national wetland reserve. The YRD is located in a temperate zone characterized by a monsoon climate. Further details about the region are shown in Table 7.1.

In this region, approximately 2,000 ha of newly formed land is created each year by sand and soil sediments carried by the river. This soil is mainly composed of fine sand and is characteristically young, with a high groundwater table, low fertility, and a tendency towards secondary salinization and desertification (Zhang 1997). The average groundwater table is generally 2–3 m, whereas it is only 0.5–1.5 m along the coastline. Given the encroachment of seawater and high evaporation rates, the mineral concentration in the soil is usually 10–40 g/L and in extreme cases 200 g/L (Fig. 7.1).

The YRD is rich in biological resources. A total of 1917 species exist, of which 50 are nationally protected and 47 internationally protected. A total of 393 plant species and varieties are found in the region, and most are salt tolerant. The natural vegetation is composed of broadleaf deciduous forest (mainly Hankow willow and weeping willow), shrubbery (mainly Chinese tamarisk), and shore coppice. Apart from indigenous species, hundreds of new salt-tolerant species have been introduced in recent years, many of which have thrived in the region (Gong 1986).

Table 7.1 The climatic conditions in the YRD

Temperature (°C)	Annual mean	12.3
	Extreme high	41.9
	Extreme low	−23.3
	Accumulated (>0 °C)	4,783.5
	Accumulated (>10 °C)	4,183
Annual radiation (MJ/m^2)		5,146–5,411
Annual sunshine time (hr)		2,682
Frost-free season (d)		210
Annual mean precipitation (mm)		542.3–842
Evaporation (mm)		1,962.1

Fig. 7.1 The YRD landscape (Photograph taken by Zhang Jianfeng)

Table 7.2 Land use in the YRD region

Available land per capita (ha)	0.5
Arable land per capita (ha)	0.19
Arable land (%)	31.7
Forestland (%)	1.7
Grassland (%)	3.8
Inhabited and mining land (%)	7.2
Roads (%)	1.5
Water bodies (rivers, reservoirs, ponds, canals, etc.) (%)	30.9
Others (%)	23.2
Total (%)	100

Among the 1917 species found in the YRD, 300 are terrestrial vertebrates (craniate), 583 are terrestrial invertebrates, 223 are terrestrial aquatic animals, and 418 are marine aquatic animals. The most important species are white stork, Chinese merganser (*Mergus squamatus*), golden eagle, white-tailed sea eagle, red-crowned crane, common crane, whooper swan, and hen harrier, along with numerous kinds of freshwater fish.

Given the low altitude and high evaporation–precipitation ratio (3.5:1), secondary salinization readily occurs in the newly formed soil (Xing 2001). The salt-affected soil has an area of 4,430 km^2. The limited forest cover and uneven distribution of vegetation lead to a weak and unstable existing ecosystem (Table 7.2). The lack of coastal protective forests and other high-quality protection systems results in increased natural disasters (storms, high tides, floods, droughts, etc.).

Precipitation in the region is low (560 mm annually); and thus, freshwater is mainly obtained from the Yellow River. However, in recent years, available water from the river has continuously decreased because of a diminishing water supply in the upper tributaries. Freshwater supply has become the limiting factor for further development.

7.2 Salinization Occurrence and Soil Features

Salinity is more a water problem than a soil problem. The YRD region is often excessively wet because of a high water table. Most cases of salinity associated with irrigation are not caused by the addition of salts from irrigation water. Instead, salinity results from the redistribution of existing salts in the soil. Canal seepage or excess irrigation may raise the water table, thus increasing the risk of salinity (Grimes and Henderson 1984).

A prerequisite for soil salinization is a free water table close enough to the soil surface to allow capillary action to lift water from the table to the soil surface. In general, if the water table is within 2 m of the soil surface, then capillary movement will carry water to the surface. However, critical depth varies with soil texture. A high water table alone is not sufficient to cause soil salinization. Rather, a high water table and evaporation that exceeds infiltration are required. That is, the salinization process is solar powered. In a semiarid climate, warm temperatures, low humidity, and wind evaporate water at the soil surface and cause salt accumulation (Nulsen 1981).

Salt accumulation occurs when the quantity of water leaving the soil surface by evaporation exceeds the quantity of water entering the soil through rainfall, runoff accumulation, or irrigation. Saline soil has a high concentration of salts that dissolve in water. Soluble salts include sodium chloride, potassium chloride, magnesium chloride, and calcium sulfate (gypsum). Sulfate salts are present in a few small areas, but chloride salts are dominant in the majority of the cases (Zhang et al. 2011).

7.3 Effects of Salt on Plant Growth

The primary effect of salt in soil is depriving plants of water. Plants need water and the nutrients dissolved in it to grow properly. The sap in plant roots contains salt that attracts water into the plant via osmotic pressure. Dissolved salts in soil increase the osmotic pressure of the soil solution (Zhang 1990), and this pressure decreases the rate at which water from the soil enters the roots (Francois et al. 1992). If the soil solution becomes too concentrated, plants will slowly starve, even though the supply of water and dissolved nutrients in soil may be more than adequate. Simply put, salt prevents water and nutrients from entering plants. A number of plants may also be adversely affected by high concentrations of certain elements, such as boron.

7.4 Strategies for Reclaiming and Ameliorating Saline Soil

7.4.1 Managing the Recharge and Discharge Areas

Water management is the key to controlling salinity, and the degree to which water movement can be managed determines the feasibility of controlling salinity. Therefore, managing recharge to cope with the discharge of groundwater is necessary.

Excess moisture that recharges groundwater can eventually cause salinity in a discharge area. Preventing excess water from accumulating and percolating deeply in the recharge area is important in controlling the salinization process at its point of origin. In several mechanisms, such as in an artesian system, the recharge area may be extensive and far from the salinity problem. In such cases, controlling recharge may not be practical.

Groundwater recharge can be controlled through the removal of excess water by improving drainage or employing cropping practices that use up more moisture. Recharge water, which is usually nonsaline and should be treated as a valuable resource rather than as a rare freshwater source in situ, should be used locally wherever possible.

Surface drainage that uses an open trench or buried pipe removes water before it has a chance to infiltrate and recharge the groundwater table. In some cases, sloughs or ponds may be consolidated to provide a water source for irrigation or other uses. Subsurface drainage may also be effective if a suitable outlet is available. However, caution should be exercised when altering natural drainage.

Continuous cropping, or growing perennial forage, uses more recharge water than summer fallow cropping. Adequate fertilization encourages better use of soil moisture regardless of crop rotation length. Deep-rooted forage crops grow well in regions with a high water table in the recharge area. Alfalfa is the deepest rooted and highest moisture-using forage crop, and thus, it is the preferred species for recharge control.

Managing the discharge area is difficult because of the presence of both excess water and salts. The objective of managing the discharge area is to establish crop growth while attempting to push salts downwards. In most dryland situations, salts are unlikely to be removed completely. However, lowering the salt concentration in the top 30 cm of the soil often improves production significantly. Even so, reclaimed saline areas always require special management practices.

7.4.2 Drainage Practice and Leaching

One method of reclaiming saline soil and restoring its productivity is by installing drainage (such as tile drains) and providing a source of leaching water (Jorgensen et al. 1992). In irrigation areas, drainage for controlling salinity may be feasible because sufficient water usually leaches excess salts out of the soil. However, in dryland areas, drainage is seldom an economically viable solution. Desalinization of dryland soil occurs very slowly upon drainage installation without extra water to leach salts out of the soil (San Joaquin Valley Drainage Program 1990). In dryland situations, improving surface drainage in recharge areas may be possible when small sloughs or other bodies of water are contributing to the problem. For example, in areas where local saline seepage has been identified, determining and draining a source of surface water that contributes to the problem may be possible.

Drainage is a controversial issue because of concerns as to where the water is going. Solving a problem in one field is pointless if it contributes to the development of a problem in an adjacent field.

7.4.3 Performance of Tillage

Salt concentrations in subsoil are frequently higher than those at the immediate soil surface; therefore, shallow tillage is generally recommended for saline soil. Deep tillage may, in many cases, bring more salts to the surface and make the problem worse (Northcote and Skene 1972).

Tillage practices should maintain as much residue as possible on the soil surface, and this objective can best be achieved by using cultivators and rod weeders instead of disk implements.

Shallow seeding is as critical in saline soil as it is in normal soil. Timeliness of seeding is particularly important for crops with small seeds, such as flax, alfalfa, sweet clover, and canola. These crops should be seeded early so that they can germinate when surface salt concentration is temporarily lower.

A number of weeds mentioned earlier are very tolerant to salt. Chemical weed control, along with tillage, is essential to grow properly established crops, particularly forage.

7.4.4 Applying Chemical Amendments

No chemical amendment that can "neutralize" salts is present in soil. The salts that are causing the problem are neutral salts (neither acid nor base). In areas where true alkali soil exists, applying elemental sulfur or gypsum can have beneficial effects. Elemental sulfur assists in reducing pH when an extremely high pH is the problem. Gypsum supplies calcium to replace sodium and improve soil structure (Zhang 1997).

7.4.5 Adding Manure and Other Fertilizers

Applying high amounts of manure (40–60 t/ha) can improve crop establishment and yields on saline soil. Manure increases the organic matter content and water-holding capacity of soil. As such, more water is available to plants even though salts are present. Manure also adds nutrients and improves the structure or tilth of soil.

Neither fertilizers nor manure directly reduces salt content. Better crop growth can result if fertilizer or manure is used with more moisture to some extent. Manure also assists the movement of water through the soil, allowing it to leach salts downwards.

For short periods, manure mulch applied on the soil surface reduces evaporation and the concentration of salts at the surface of the soil (Peck 1978). Other forms of mulch, such as straw, probably perform the same function. When mulch is present and significant precipitation occurs, leaching of salts from the surface layer may even be possible.

On moderately saline soil, using phosphate fertilizers with cereal grain seeds has produced excellent yield increases. Nitrogen fertilizers do not usually provide high yield responses in areas most seriously affected by salinity. However, if salinity is scattered within a field, then nitrogen fertilizer should be applied according to soil test recommendations for nonsaline areas. Interpreting nitrogen soil tests is often complicated by the presence of salts; and thus, areas that are obviously saline should not be sampled. Special attention should be paid to the location of available nitrogen in the soil profile under saline soil conditions. Saline subsoil usually contains a high concentration of nitrate–nitrogen that may not be useful to plants.

7.4.6 Planting Crops and Forage

Planting crops is the main management practice for combating soil salinity problems in discharge areas. Leaving the soil bare promotes evaporation and salt accumulation on the surface (Zhang 2001). The tolerance levels of various annual herbs, forage, and vegetable crops are different. Winter wheat, sunflower, and soybeans are mostly grown when salinity is not very high.

If annual crops cannot maintain economic yields on saline land, then perennial forage production is the only management practice that can be recommended (Webb and Price 1994). Mixtures of grass and legumes are seeded so that some plant species will be established even in the most saline areas, and the more valuable species will grow on less saline portions of the field.

Legumes with a certain degree of salt tolerance, for example, sweet clover and alfalfa, are recommended for improving soil quality. Alfalfa and sweet clover are the backbone of salinity reclamation. Both species fix their own nitrogen when properly inoculated and are capable of prolific growth under good moisture conditions. Sweet clover is established easily and grows vigorously in saline conditions. This legume is a proven green manure crop for improving soil. One drawback of sweet clover is its short life span (biennial). Alfalfa is a long-lived perennial legume that has a deep taproot system. Once established, this species provides excellent production for many years.

Common salt-tolerant grass species, for example, Altai wild ryegrass, tall wheat-grass, Russian wild ryegrass, and slender wheatgrass, exhibit relatively equal salt tolerance levels, but play different roles in managing saline soil. Slender wheatgrass is the most easily established, followed by Russian wild ryegrass, tall wheatgrass, and Altai wild ryegrass. Slender wheatgrass is short-lived and best suited to short-term rotations. Under dry conditions, slender wheatgrass is more salt tolerant than tall wheatgrass. Altai wild ryegrass has very limited creeping ability and is most useful in pastures. This type of grass has a high degree of tolerance to soil salinity and has the deepest rooting system among grass species; however, this ryegrass is slow to establish.

Russian wild ryegrass is similar to Altai wild ryegrass in terms of its salt tolerance level, and it is most useful in pastures. However, Russian wild ryegrass is not as deep-rooted as Altai. Russian wild ryegrass often establishes very slowly, and thus, patience is required when establishing this crop. Tall wheatgrass is the most salt tolerant and produces the largest yield of forage hay. This species is also the most tolerant to flooding and extremely wet conditions and as such is most useful in areas seriously affected by salinity. Although tall wheatgrass is the most salt tolerant, it is not the most palatable or nutritious livestock feed; thus, mixing with other forage species is required in any feeding program. Given their high tolerance to salt, these accepted salt-tolerant grass species should be included in the seeding mixture for saline land.

7.4.7 *Establishing Forests*

One of the main strategies for coping with saline soil problems is lowering the groundwater table (Zhang et al. 2011), and trees can help control groundwater levels in two ways: by decreasing recharge and by increasing discharge. Most groundwater recharge is supplied by rainfall (except in the case of irrigated-land salinity), and more recharge reaches groundwater when trees are cleared. This situation happens because trees develop extensive root systems that trap water, which is then used for tree growth or returned to the atmosphere by evaporation and transpiration. Scientists estimate that the amount of water percolating below the root zone of crops and pastures can be 10–100 times the amount of water percolating below trees.

Trees also play a role in discharging groundwater (Fig. 7.2). The roots of many tree species reach deep into the ground, often making contact with the water table. During droughts, such trees use this water to survive and continue growing (Zhang et al. 2010). By contrast, annual crops planted by farmers on cleared land usually do not have deep roots. Moreover, numerous crops only grow during winter, spring, and early summer, whereas most trees require water throughout the year.

Tree species suitable for the YRD region include Chinese ash, tree of heaven, Chinese sumac, Chinaberry tree, Siberian elm, and black locust. False indigo, *Amorpha*, and such trees can be mixed within the rows. Different tree species are best suited for various land uses. Naturally, tree species must be selected with respect to in situ soil conditions (Zhang et al. 2010) to establish shelterbelts for livestock farms. These species may include Siberian elm, black locust, Chinese honey locust,

Fig. 7.2 Trees are planted to increase vegetation coverage (Photograph taken by Zhang Jianfeng)

and panicled golden rain tree. In land where secondary salinization has occurred, the Taungya system should be adopted. Tree species can be jujube, *Ginkgo biloba*, pear, and apricot, and the space required is 50–80 × 3–6 m. Possible intercropping crops can include groundnut and watermelon during the early stage and soybean and forage grass (herbage) in later stages. A mixed forest is the first priority in establishing a timber production base. The combination can be black locust + tree of heaven, Chinese ash + false indigo Amorpha, false indigo Amorpha + poplar, Japanese pagoda tree + Siberian elm, and Chinese honey locust + false indigo Amorpha.

Salinity in the YRD region is a serious problem that hinders regional sustainable development. The main cause of the problem is dryness, over-evaporation leading to a rising groundwater table, and consequently, salt accumulating in certain areas. Saline soil has a high concentration of soluble salts that reduces the ability of plants to absorb water and nutrients. Salinity can be controlled in certain situations once the specific causes are understood. Furthermore, salinity can be reduced by appropriate management practices. In the recharge area, improved drainage and water-efficient crop management practices can reduce the amount of water entering the groundwater system. In the discharge area, where salinity is observed, establishing plant growth is important. Suitable cultivation systems, including leaching, tillage, and drainage, are practical strategies for ameliorating salt-affected soil. For a number of specific situations, applying amendments to improve soil structure and properties is effective. The natural addition of manure and other organic or chemical fertilizers increases production by enhancing soil fertilization, although soil salinity is not directly reduced. Planting salt-tolerant crops and forage and establishing forests as a long-term management strategy are the optimal choices for reclaiming saline soil. However, the process of selecting suitable species is critical and time-consuming.

References

Francois LE, Donovan TJ, Maas EV (1992) Yield, vegetative growth, and fiber length of kenaf grown on saline soil. Agron J 84:592–598

Gong HZ (1986) Silviculture in saline soil. China Forestry Press, Beijing

Grimes DW, Henderson DW (1984) Developing the resource potential of a shallow water table. University of California, Davis, p 39

Jorgensen GS, Solomon KH, Cervinka V (1992) Agroforestry systems for on farm drain water management. In: Proceeding America society of Agricultural. Engineers, sixth international drainage symposium, Nashville, 13–15 Dec, pp 484–490

Northcote KH, Skene JKM (1972) Australian soils with saline and sodic properties, vol 27, Soil publication. CSIRO, Australia

Nulsen RA (1981) Critical depth to groundwater in non-irrigated situations. Aust J Soil Res 19:83–86

Pavelic P, Dillon PJ, Narayan KA et al (1997) Integrated groundwater flow and agronomic modelling for management of dryland salinity of a coastal plain in southern Australia. Agric Water Manag 35:75–83

Peck AJ (1978) Note on the role of a shallow aquifer in dryland salinity. Aust J Soil Res 16:237–240

San Joaquin Valley Drainage Program (1990) A management plan for agricultural subsurface drainage and related problems on the westside San Joaquin Valley. Final report, Imhoff E Program Manager, Sacramento. p 183

Webb A, Price R (1994) Procedures for economic assessment of management options for dryland salinity. Land and Water Resources Research and Development Corporation, Occasional paper no 06/94

Xing SJ (2001) Study on salinity tolerance and planting techniques of *Nitraria* in worst salt affected soil. In: Liu MY (ed) The proceedings of international symposium on halophyte utilization and regional sustainable development of agriculture. Meteorology press, Beijing

Zhang JF (1990) Study on plantlets cultivation of new variety 74009 of white elm. M.Sc. Dissertation, Shandong Agricultural University

Zhang JF (1997) Review of saline soil utilization. J Shandong ForSci Technol 3:5–8

Zhang JF (2001) Principles and practice of afforestation in saline soil in China. In: Liu MY (ed) The proceedings of international symposium on halophyte utilization and regional sustainable development of agriculture. Meteorology press, Beijing

Zhang JF, Chen GC, Xing SJ et al (2010) Carbon sequestration of black locust forests in the Yellow River delta region, China. Int J Sustain Dev World Ecol 17(6):475–480

Zhang JF, Chen GC, Xing SJ (2011) Water shortages and countermeasures for sustainable utilisation in the context of climate change in the Yellow River Delta region, China. Int J Sustain Dev World Ecol 18(2):177–185

Part III
Saline Soil Utilization for Biomass Production

Chapter 8
Biomass Production in Saline Soil

Abstract At present, land use is relatively varied because of different climates and soil types. On nonarable land, the focus should be conserving and improving pastures from the perspective of soil improvement and sustainability, particularly for indigenous vegetations. On arable land, emphasis should be placed on establishing windbreaker forests. In selecting species, both economic and ecological benefits should be considered. China is home to more than 4 million ha of oil plants; 154 kinds of energy trees can produce seeds containing more than 40 % of oil, with a total production of seeds reaching 5 million tons. Another 57 million ha of wasteland is available and suitable for planting trees for the production of forest-based bioenergy. The potential for forest-based bioenergy in the country is thus equivalent to 200 million tons of coal, the use of which will reduce consumption of fossil energy by 10 %. Ultimately, biomass can be used in carbon-negative energy systems, thus further reducing carbon dioxide emissions. According to the State Forestry Administration, China aims to establish 13.33 million ha of forests by 2020 to produce liquid and solid biofuels for power generation and transport. The target hectarage will rise from a planned 833,333 ha in 2010 and will provide enough biomass to produce more than 6 million tons of biodiesel each year and power generators with a capacity of 15 GW. These energy forests will contribute significantly to the already successful attempts by China to reduce desertification. Undoubtedly, saline land exploitation will play a significant role in this context.

Keywords Biomass production • Renewable energy • Saline soil

China is currently undergoing a phase of rapid industrialization and integration into the global economy. However, this progress comes with a high price that exerts tremendous strain on the environment through the extensive use of fossil fuels and other natural resources. The government and the people are experiencing increasingly higher pressure, particularly with the rising price of petroleum in

J. Zhang, *Coastal Saline Soil Rehabilitation and Utilization Based on Forestry Approaches in China*, DOI 10.1007/978-3-642-39915-2_8,

the global market. Hence, China has been anxious to boost its use of alternative energy sources in recent years because its heavy reliance on burning coal to satisfy energy requirements has led to serious environmental problems. Consequently, one of the alternatives found is to develop bioenergy (Zhang et al. 2009a). With its vast territory and diverse geographical regions, China has a huge stock of biomass resources from agricultural and forest residues, as well as vast areas of salt-affected land that can be used for bioenergy development by cultivating salt-tolerant plants. The harvests from such plants can be used as raw materials for bioenergy production, such as in small and decentralized electricity and heat generation, household applications, and biofuels (Fung et al. 2002).

Bioenergy development has become a priority for the government. The current focus is on electricity generation from surplus agricultural residues, which are produced at an estimated level of 200 million tons yearly. According to the State Forestry Administration, China aims to establish 13.33 million ha of forests by 2020 to produce liquid and solid biofuels for power generation and transport (Lei 2002). The target hectarage will rise from a planned 833,333 ha in 2010 and will provide enough biomass to produce more than 6 million tons of biodiesel each year and power generators with a capacity of 15 GW. These energy forests will contribute significantly to the already successful attempts of China to reduce desertification. Undoubtedly, saline land exploitation will play a significant role in this context.

Moreover, power plants that use biomass can help improve the quality of life of people living in salt-affected areas, increase agricultural land resources, develop the local ecological environment, and promote sustainable development in the region (Stokes 1992; Bob et al. 2003).

8.1 Saline Soil Resources in China

Saline soil is an important land resource extensively distributed across the Earth, including Euroasia, Africa, and western North America (Szabolcs 1989). Saline soil has an approximate total area of more than 1 billion ha, covering up to 25 % of the land area on the planet (Kovda 1983). Approximately 27 million ha of saline–alkali land is found in China, where the distribution of saline soil varies significantly because of the huge land area and diverse climatology of the country. Salt ingredients and the salinization process also contribute to the variation (Zhang 2008).

The North China Plain (NCP), a major food production area, is one region in the country with a serious saline soil problem. For example, during the early 1950s, the NCP had more than 2 million ha of saline soil. By the end of the 1950s, the coverage of affected areas had increased to 4 million ha. However, in the mid-1980s, the number returned to 2 million ha. This overall decrease in the extent of saline can largely be attributed to changes in irrigation practices and an associated drop in

Fig. 8.1 Saline land at Bohai Gulf (Photograph taken by Zhang Jianfeng)

groundwater table depth in the region. Areas where the extent of salinization is still considered serious include the lower elevated plains of the Bohai Gulf, the four lakes in the south of Shandong Province, and the Yellow River irrigation area in Henan Province (Fig. 8.1).

Saline soils on the NCP are characterized by accumulated salts near the surface, which is a characteristic that can be attributed to a variety of reasons. Given the critical depth of saline groundwater and the intense evaporation rate, salts have progressively moved or have been wicked from saline groundwater and lower-lying soil. Salinity has also extended from lower topographic sites to higher sites because of uneven microtopography. The level of salinity in soil is closely correlated with crop development and yield because cultivation management, seed emergence, and crop growth occur in the plow layer (the major rooting depth). Thus, the salt content of the soil surface (0–20 cm) can be used to categorize the extent and nature of soil salinization and to provide a sound basis for rehabilitating saline–sodic soil. Good farming practices and appropriate biological approaches enable redistribution and leaching of salt out of the soil.

In general, soil salinity in the top layer is approximately 0.4–3 %, which results in poor structural properties and fertility (Zhang et al. 2008). Without improving management methods, increasing agricultural production in these lands can be difficult. Exploiting saline land by cultivating salt-tolerant plants for biomass production is highly feasible.

8.2 Flora Resources in Saline Soil for Biomass Production

More than 393 plant species and varieties grow on saline land, most of which are salt tolerant. Natural woody vegetation includes broadleaf deciduous forest (mainly Hankow willow and weeping willow), shrubbery (mainly Chinese tamarisk), and shore coppice. Herbaceous vegetation includes *Pelargonium hortorum* Bailey., *Rubia cordifolia* Linn., *Digitaria sanguinalis* (Linn.) Scop., *Erysimum cheiranthoides* Linn., *Salsola collina* Pall., *Hippochaete ramosissimun* Desf., *Viola philippica* Cav., *Cynosurus viridis* (Linn.) Beauv., *Imperata cylindrical* (Linn.) Beauv., *Meconopsis kanashiroi* Kitam., *Afzelia nigrum* Linn., *Meconopsis hedinii* Ostenf. et Pauls., and *Argyranthemum coronarium* Linn (Fig. 8.2).

Apart from the indigenous species, hundreds of new salt- and wind-tolerant species have been introduced in recent years, many of which have thrived in the region (Gupta et al. 1987; Marcar et al. 1999; Mainguet 1999). These species include Japanese black pine (*Pinus thunbergiana*), eastern red cedar (*Juniperus virginiana*), Hollywood juniper (*Juniperus chinensis*), live oak (*Quercus virginiana*), yaupon (*Ilex vomitoria*), salt cedars (*Tamarix* spp.) and shrubs such as salt shrub (*Baccharis halimifolia*), dwarf yaupons (*Ilex vomitoria* "stokes dwarf" and other cultivars), thorny elaeagnus (*Elaeagnus pungens*), butcher's broom (*Ruscus aculeatus*), wax myrtle (*Myrica cerifera*), northern bayberry (*Myrica pensylvanica*), Scotch broom (*Cytisus scoparius*), pittosporum (*Pittosporum tobira*), Spanish bayonet (*Yucca*

Fig. 8.2 Wild plants growing on the coast of Shanghai (Photograph taken by Zhang Jianfeng)

gloriosa), Yedda hawthorn (*Rhaphiolepis umbellata*), century plant (*Agave americana*), Laurustinus (*Viburnum tinus*), Osmanthus (*Osmanthus* spp.), and big-leaf hydrangea (*Hydrangea macrophylla*). Annuals include ageratum (*Ageratum houstonianum*), blanket flower (*Gaillardia pulchella*), moss rose (*Portulaca oleracea*), verbena (*Verbena xhybrida*), Mexican heather (*Cuphea hyssopifolia*), dusty miller (*Senecio cineraria*), lantanas (*Lantana* spp.), sweet alyssum (*Lobularia maritima*), ivy geranium (*Pelargonium peltatum*), and calendula (*Calendula officinalis*).

8.3 Strategies for Reclaiming and Developing Saline Soil

At present, land use is relatively varied because of different climates and soil types. On nonarable land, the focus should be conserving and improving pastures from the perspective of soil improvement and sustainability, particularly for indigenous vegetations (Zhang 2004). On arable land, emphasis should be placed on establishing windbreaker forests. In selecting species, both economic and ecological benefits should be considered (Malcolm and Sumner 1998; Qureshi and Barrett-Lennard 1998). Selected species may include Chinese ash, white mulberry, Chinese wolfberry, sarch, and peach. In a reed coenosium site, the task is to improve its inhabitant conditions, increase its productivity, and enhance its protective functions.

1. *Establishing shelterbelts for crops*. The recommended area of the cell is 6.7–10.0 ha. Species may include Chinese ash, tree of heaven, Chinese sumac, China berry tree, Siberian elm, and black locust. False indigo Amorpha and sarch can be mixed within rows.
2. *Establishing shelterbelts for livestock farms*. The recommended area of the cell is 13.3 ha. Species may include Siberian elm, black locust, Chinese honey locust, and panicled golden rain tree.
3. *Implementing agroforestry systems*. In land where secondary salinization has occurred, the Taungya system should be adopted. Tree species can include jujube, *Ginkgo biloba*, pear, and apricot. The spacing should be 50–80 × 3–6 m. Intercropping crops can include groundnut and watermelon in the early stage and soybean and forage grass (herbage) in later stages.
4. *Establishing a timber production base*. A mixed forest is the first priority. Combinations can include black locust + tree of heaven, Chinese ash + false indigo Amorpha, false indigo Amorpha + poplar, Japanese pagoda tree + Siberian elm, and Chinese honey locust + false indigo Amorpha.
5. *Improving existing orchards*. Priority is given to unique and rare species with high quality and efficiency.

Planting and protecting vegetation in saline land can benefit farmers by improving their health and living conditions, reducing fossil fuel usage, creating jobs, and generating income (Zhang 2002; Zhang and Li 2002). In addition to other environmental and social benefits, the same amount of investment in household biomass use applied in biomass power plants can generate 5–10 times more local jobs for rural residents and 5–9 times more income for small companies (Zhang 2003; Zhang et al. 2009b).

China is home to more than 4 million ha of oil plants; 154 kinds of energy trees can produce seeds containing more than 40 % of oil, with a total production of seeds reaching 5 million tons. Another 57 million ha of wasteland is available and suitable for planting trees for the production of forest-based bioenergy. The potential for forest-based bioenergy in the country is thus equivalent to 200 million tons of coal, the use of which will reduce consumption of fossil energy by 10 %. Ultimately, biomass can be used in carbon-negative energy systems, thus further reducing carbon dioxide emissions.

The aim for China is to increase its hectarage of trees suitable for providing feedstock for power plants and biodiesel manufacturers by 16 times during the period 2010–2020. By integrating the reduction of greenhouse gas emissions into the sustainable development of rural energy systems, China can set an example for other biomass-rich developing countries as they strive to achieve the combined benefits of social development and environmental protection.

References

Bob R, Jeff P, Dennis M et al (2003) A strategic assessment of forest biomass and fuel reduction treatments in western states. U.S. Department of Agriculture, Forest Service, Research and Development, Washington, DC, p iii, 18 pages

Fung PYH, Kirschbaum MUF, Raison RJ et al (2002) The potential for bioenergy production from Australian forests, its contribution to national greenhouse targets and recent developments in conversion processes. Biomass Bioenergy 22:223–236

Gupta GN, Mohan S, Prasad KG (1987) Salt-tolerance of selected tree seedlings. J Trop For 3(3):217–227

Kovda VA (1983) Loss of productive land due to salinazation. Ambio 7(2):91–93

Lei JF (2002) The operation of forest resources in China. China ForEcon 4:3–5

Mainguet M (1999) Aridity drought and human development. Springer, Berlin/Heidelberg

Malcolm E, Sumner RN (1998) Sodic soils-distribution, properties, management, and environmental consequences. Oxford University Press, New York

Marcar N, Ismail S, Hossain A (1999) Trees, shrubs and grasses for saltlands. Australian Centre for International Agricultural Research, Canberra

Qureshi RH, Barrett-Lennard EG (1998) Saline agriculture for irrigated land in Pakistan: a handbook. Australian Centre for International Agricultural Research, Canberra

Stokes BJ (1992) Harvesting small trees and forest residues. Biomass Bioenergy 2(1):131–147

Szabolcs I (1989) Salt-affected soils. CRC Press, Boca Raton

Zhang JF (2002) Research on saline soil amelioration and forestation techniques. J Northeast ForUniv 30(6):124–129

Zhang JF (2003) Advances of research on the mechanism of plant salinity tolerance and breeding of salt-tolerant plants. World ForRes 16(2):16–22

Zhang JF (2004) Agroforestry and its application in amelioration of saline soils in eastern China coastal region. For Study China 6(2):27–33

Zhang JF (2008) The principles and techniques on ecological rehabilitation of saline soils. China Forestry Press, Beijing, pp 24–46

Zhang JF, Li XF (2002) Strategies to reclaim and ameliorate saline soil in the Yellow River Delta Region. In: Chi Chang, Brian D (eds) International conference on environmentally sustainable agriculture for dry areas for the 3rd millennium proceedings. Dobing Enterprises, Canada, pp 264–269

Zhang JF, Jiang JM, Xing SJ (2008) Planting techniques of *Tamarix chinensis* and its effect on saline soil remediation. In: Proceedings of the 2nd international conference on bioinformatics and biomedical engineering. Shanghai, China

Zhang JF, Jiang JM, Shan QH (2009a) Research on the black locust (*Robinia pseudoacacia*) forest as biomass energy resources in Yellow River delta region. In: Proceedings of Asia-Pacific power and energy engineering conference. Wuhan, China

Zhang JF, Zhang DS, Fan BM, Shan QH et al (2009b) Characteristics of salt affected soil and its amelioration by trees. Chin For Sci Technol 8(1):42–45

Chapter 9
Potential and Availability of Forest Biomass Energy Production in the East China Sea Regions

Abstract In the East China Sea region, tree planting can help manage environmental problems and provide alternative energy sources by significantly improving circumstances and producing renewable biomass energy. Biomass plantations provide an effective means of reducing greenhouse gas emissions and offer benefits to rural residents by diversifying their sources of income through providing multipurpose products. A larger area for fuelwood plantations is required to improve the rural energy situation and reduce the pressure from fuel gathering. The incentives in maintaining such plantations depend on the land areas and relative fuel prices in the locality. Fuelwood production should ideally be included in an integrated watershed development program, in which forestry is combined with crops and pasture on appropriate soil and slopes to maximize the conservation of soil and moisture. China needs to reforest almost 5 million ha per year for the next 25 years and increase the amount of forested land to approximately 20 % for this program to have a significant effect. An effective means of reducing greenhouse gases is using wood and other biomass and their residues as substitutes for coal. This practice increases carbon sequestration. Apart from providing sustainable energy in gaseous, liquid, and solid forms, bioenergy can create new job opportunities, improve rural infrastructure, and enhance the local and global environment.

Keywords Biomass energy • East China Sea • Forestry

The world has recently embarked on a new round of biomass energy development and sustainable utilization of global resources as a result of continuous oil price hikes, worsening environmental pollution, and global climate change (Li et al. 2004; Zhang et al. 2010). Less arable land is one of the basic national conditions in China. Numerous mountains, beaches, saline–alkali soil, and other marginal lands are found throughout the country. Given the complexity of the factors involved in climate and soil conditions, using marginal grain-growing land for cultivating

J. Zhang, *Coastal Saline Soil Rehabilitation and Utilization Based on Forestry Approaches in China*, DOI 10.1007/978-3-642-39915-2_9,

energy crops to generate biomass energy is difficult. Moreover, providing sufficient raw materials, such as sweet sorghum, cassava, and other non-grain crops, may also be a problem. In this context, shorelines play an important role in the field.

9.1 Importance of Developing the East China Sea Region

A coastal zone is the transition zone between terrestrial and freshwater/marine ecosystems. Extending for more than 18,000 km, the coastline in China is bordered by the Bohai Sea, Yellow Sea, East China Sea, and South China Sea. With socioeconomic development, an increasing number of people are moving to the east in search of better opportunities. Hence, the East China Sea region, which is located between 23°00′–33°10′ E and 117°11′–131°00′ N, is becoming increasingly significant. For example, coastal Shanghai is located in the coastal areas along the Yangtze River. The Yangtze River Delta is an important component of the unique geographical advantage of this region, strategically positioned on land where resources are both rich and abundant. Today, expediting the development of the Yangtze River Delta region is crucial. Developments are focusing on upgrading industry, enhancing overall strength, improving the productivity of national coastal areas, and promoting the development of the central and western regions through exchange and cooperation not only within China but also between China and Central/Northeast Asian counties and Europe. At the same time, these developments also aim to strengthen coastal ports; to improve water conservation, transportation and energy infrastructure, power grids, and other major buildings; as well as to constantly enhance the ability to support regional development.

9.2 Main Constraints in the Region

In the past two decades, the environment of the East China Sea region has faced huge stresses resulting from anthropogenic activities and population growth in the Yangtze River drainage basin and coastal areas. Numerous pollutants from land-based sources, such as sewage, oil hydrocarbons, sediments, nutrients, pesticides, litter, marine debris, and toxic waste, enter the sea with river water and other runoff from land. These pollutants threaten coastal and marine ecosystems, as well as the health of coastal inhabitants, by limiting phytoplankton growth, increasing the mortality of fish and benthos, enhancing eutrophication, promoting the occurrence of red tide, decreasing fishery yields, and causing other negative nonreversible changes in ecosystem health. At present, contamination from land runoff is the most significant contributor to coastal environmental pollution. The dominant pollutants are inorganic nitrogen, phosphate, oil hydrocarbons, organic matter, and heavy metals. Although the volume of runoff from industrial effluents into coastal water has decreased since 1999, municipal sewage and other runoff from land have exhibited an increasing trend (Li and Daler 2004). Another crucial issue is the

Fig. 9.1 Wind power in the coastal area of Fuzhou, Fujian Province (Photograph taken by Zhang Jianfeng)

shortage of power in the region, which has become ever more serious with the increasing population (Fig. 9.1).

Tree planting helps manage environmental problems and provide alternative energy sources by significantly improving environmental conditions and producing renewable biomass energy in the region. Establishing biomass plantations provides an effective means of reducing greenhouse gas emissions, as well as offers rural residents a diverse source of income through providing multipurpose products. A larger area for fuelwood plantations is required to improve the rural energy situation and reduce the pressure from fuel gathering. The incentives in maintaining such plantations depend on the land areas and relative fuel prices in the locality. Fuelwood production should ideally be included in an integrated watershed development program, in which forestry is combined with crops and pasture on appropriate soil and slopes to maximize conservation of soil and moisture. China needs to reforest almost 5 million ha per year for the next 25 years and increase the amount of forested land to approximately 20 % for this program to have a significant effect. An effective means of reducing greenhouse gases is using wood and other biomass and their residues as substitutes for coal. This practice increases carbon sequestration. Apart from providing sustainable energy in gaseous, liquid, and solid forms, bioenergy can also create new job opportunities, improve rural infrastructure, and enhance the local and global environment.

9.3 Environmental Values of Coastal Areas

Coastal areas provide a huge number of environmental products and services. Their ecological importance has been widely accepted regardless of which approach is considered, that is, utilitarian or environmentalist. The main characteristic of coastal ecosystems is their dynamic nature resulting from the transfer of matter, energy, and organisms between land and sea, under the influence of primary driving forces, such as weather and climate, rises in sea level, and tides (Bowen and Riley 2003).

Given that marine, estuarine, and coastal areas obtain their water both from land and ocean upwelling, these regions often have nutrient-rich water with high biological productivity. Moreover, coastal areas often have critical terrestrial and aquatic habitats, including estuarine areas, coral reefs, coastal mangrove forests, coastal wetlands, tidal flats, and sea grass beds. These regions comprise unique ecosystems with rich biodiversity. Approximately 90 % of the fish products in the world are dependent on coastal areas at some time during their life cycle (Hildebrand 2002). These areas support numerous migratory and nonmigratory waterfowl and shorebirds. Maintaining biodiversity is generally accepted as imperative in various societies. The physical features of coastal ecosystems, such as mangrove belts, can mitigate the effect of both natural disasters (e.g., storm-tide surges, shoreline retreat, floods) and natural processes (e.g., coastal erosion, land accretion, damage from waves, wind action).

Even if coastal ecosystems are not unique biological systems, their location at the land–sea interface makes them valuable from the recreational and aesthetic perspectives. These ecosystems support tourism activities and provide attractive sites for residential areas and industrial developments.

9.4 Economic Values of Coastal Regions

Historically, coastal areas have been a major habitat for humans because of their favorable biophysical and climatic conditions, along with the ease in communication and navigation. Harbors, which have always been the nuclei of development in the past, have gained increasing importance today. Traditional resource-based activities, such as coastal fisheries, aquaculture, agriculture, and forestry, are now competing with other activities, such as shipping, industry, and tourism. From the social and economic perspective, the importance of coastal areas is indubitable because these resources support major economic and subsistence activities. Coastal resources allow the development of agriculture, fisheries, forestry, mining, oil and gas extraction, marine transport (80 % of the world's cargo is currently being shipped across oceans and coasts), and tourism (Turner 2000). Many of the most productive agricultural lands in the world are located in river deltas and coastal plains.

In the future, dependence on coastal resources is likely to remain strong under the conditions of urbanization and economic diversification (mostly in developing countries). Industrial development often entails processing of agricultural, fishery, and forestry products, along with oil refining and textile manufacturing. These diverse economic activities are often dependent on coastal resources. An increase in economic diversification causes component sectors to be more interdependent, while conflicts over natural resources and the environment tend to develop. These conflicts exert tremendous pressure on fragile coastal environments. Complicated environmental issues are already threatening several areas.

9.5 The Necessity and Feasibility of Developing Bioenergy

Energy consumption in China has been increasing in recent years with rapid population growth and economic development. Compared with limited fossil energy resources, biomass energy is an ideal solution to the energy crisis (Yamamoto and Matsumura 2005; Zhang et al. 2009). First, biomass is the only renewable carbon source on Earth. Carbon is the basic source of the current system of fuel and bulk chemicals, and a basis for other renewable energy sources is not directly available. Second, biomass can be directly converted into liquid form for developing renewable biomass liquid fuel. This alternative source of energy can help lessen the country's dependence on foreign oil-based liquid fuels. After the depletion of fossil energy sources in the future, liquid fuel can still be an indispensable and important energy source directly provided by biomass (Serio et al. 2007).

However, the development of biomass energy in China faces a number of bottlenecks, including scarcity and low quality of raw materials, collection and transformation difficulties, catalytic biomass conversion efficiency, low energy and water consumption, a difficult biological transformation process from low-cost to large-scale amplification, poor quality bioenergy end products, and a lack of product standards.

China has approximately 100 million surplus laborers in rural areas and a vast wasteland in coastal regions. If these resources can be combined to plant trees, income can be generated (forest products and bioenergy) and the environment can be significantly improved. In addition, a lack of funds is not a problem in China because Chinese people deposit trillions of yuan into bank accounts at a negative interest rate. This huge amount of money produces less profit compared to when it is invested in tree plantations. Capital does not work well with combined labor and land where trees are planted because of the weakness of institutional arrangements, including insecure ownership, gradual capital recovery, and high risks in the plantation industry (pests, fire, theft, etc.). Without doubt, developing a legal system could improve ownership security, establish a market for young trees, and ensure that the business can withstand risks. Such a policy can also help relieve poverty in remote areas.

Therefore, the development of the bioenergy industry in China should adhere to the principles of eco-development. Establishing bioenergy eco-technology and eco-industrial systems should not compete with the production of food, oil, grains, and water to meet the needs of the people. Nature and development should work harmoniously in producing bioenergy (Zhang 2004). In practice, we should uphold resources, technology, and product diversification and vigorously develop energy plants, algae, and microorganisms to enhance the diversity of producing bioenergy resources, to develop thermochemical conversion, and to facilitate catalytic conversion of green chemistry, industrial biotechnology, and so on, with features of the core technology. We should vigorously develop biomass power generation, biogas, bioliquid fuel, and other forms of bioenergy end products (Xiang 1996).

Finally, China possesses a huge volume of waste biomass resources and inappropriate marginal farming land. Developing the biomass energy industry addresses the issues of waste resource reduction, harmful resource utilization, and sufficient energy provision. Moreover, this practice can improve the ecological environment, develop related industries, and promote the formation of a new socialist countryside.

9.6 Functions of Energy Plants

Forestry-based biomass energy sources in China are rich in raw materials. Approximately 4000 species of energy plants can be found throughout the country. Thus, the development potential for forest biomass energy is vast, for example, more than 10 species of oil plants have been identified. Moreover, the seed oil content rate of more than 40 % of 150 kinds of plants can be cultivated with the large-scale use of trees and shrubs. At present, more established development and utilization processes of biodiesel are being applied to jatropha tree, *Pistacia chinensis*, *Betula* trees, shiny fruit, tung oil trees, and other species, such as *Sapium sebiferum*, *Metroxylon sagu*, *Manihot esculenta* Crantz, *Tetraena mongolica*, *Quercus mongolicus*, *Lithocarpus dealbatus*, and *Cinnamomum pauciflorum* Nees.

These trees can be planted in coastal areas as raw materials for producing bioenergy. They can also act as buffers for protecting the environment (Zhang 2004). Buffers are vital ecological resources serving numerous roles. Their ability to remove nitrogen from groundwater and surface water is one such important function. Although buffers have been widely accepted by the scientific community as effective pollutant traps, debates are still ongoing regarding the efficiency of the removal rate. The majority of nitrogen pollutants enter streams and rivers through groundwater (Zhang 2004). Buffers take in nitrates as a plant nutrient, reduce it into nitrogen gas, and release the harmless gas into the atmosphere, thereby limiting nitrate concentrations in surface water. Riparian buffers play a critical role in protecting the health of streams.

Fig. 9.2 Field survey at a beach in Shanghai (Photograph taken by Xi Jinbiao)

Riparian forest buffers can reduce 30–90 % of nutrients and sediments. Other benefits of buffers include leaf food for fish and other aquatic organisms when falling into streams, fish/wildlife habitat, nutrient uptake stored in woody vegetation through denitrification, canopy/shade that keeps water cool and rich in dissolved oxygen, and runoff filter that slows down water velocity—infiltration 10–15 times more than grass turf and 40 times more than a plowed field.

The biggest limiting factor with biomass energy production in China is probably finding land to grow biomass needed to produce energy. Using land intended for food production or clearing forestland is clearly not a good long-term solution (Shi 2004). Fortunately, almost 1 million ha of saline–alkali soil, sand, mining sites, oil reclamation land, and other inappropriate developments in marginal agricultural land can become a base for developing biomass energy (Miu and Wu 2003). This process is an optimum means of using these abandoned lands, such as a beach enclosed by tidal and in a coastal region (see Fig. 9.2). China has 175 million ha of existing forest areas, with a forest accumulation of 12.456 billion m^3, along with barren mountains of more than 5,400 million ha of wasteland, and almost 1 million ha of saline–alkali soil, sand, and mining and oil reclamation sites. Using these resources fully for developing biomass energy can improve the energy structure in China, reduce its dependence on fossil energy, and protect national energy security—all of which are extremely significant (Azam et al. 2005; Xu 2003).

References

Azam MM, Waris A, Nahar NM (2005) Prospects and potential of fatty acid methyl esters of some non-traditional seed oils for use as biodiesel in India. Biomass Bioenergy 29:293–302

Bowen RE, Riley C (2003) Socioeconomic indicators and integrated coastal management. Ocean Coastal Manag 46:299–312

Hildebrand L (2002) Integrated coastal management: lessons learned and challenges ahead. Discussion document for managing shared water, Coastal Zone Canada Association, Hamilton

Li DJ, Daler D (2004) Ocean pollution from land-based sources: East China Sea, China. Ambio 33(1–2):107–113

Li GL, Wang YH, Yang JT (2004) Utilization condition and outlook of biomass energy in China. J Henan Agric Univ 38(1):100–104

Miu XL, Wu QY (2003) Exploitation of biomass renewable energy sources of microalgae. Renew Energy 3:13–16

Serio MD, Cozzolino M, Tesser R et al (2007) Vanadyl phosphate catalysts in biodiesel production. Appl Catal A Gen 320(22):1–7

Shi LS (2004) Analyzed of Chinese energy status and programming of renewable energy. Renew Energy 5:1–4

Turner KR (2000) Integrating natural and socio-economic science in coastal management. J Mar Syst 25:447–460

Xiang S (1996) The study and development on plant fuel. Sci Chin 12:43–45

Xu YY (2003) A novel enzymatic route for biodiesel production from renewable oils in a solvent-free medium. Biotechnol Lett 25:1239–1241

Yamamoto H, Matsumura Y (2005) Evaluation of supply potential of enemy crops in Japan considering eases of improvement of crop productivity. Biomass Bioenergy 29(5):355–359

Zhang JF (2004) Agroforestry and its application in amelioration of saline soils in eastern China coastal region. For Study China 6(2):27–33

Zhang JF, Sun QX, Zhou JX et al (2009) Biomass production of poplar plantation ecosystem in Yangtze River Beach Land. Energy Power Eng 1(2):81–84

Zhang JF, Chen GC, Shan QH (2010) Potential and future prospects of biomass production in saline soils. In: Proceedings of Asia-Pacific power and energy engineering conference. Chengdu, China

Part IV
Wetland Degradation and Water Shortage in the Yellow River Delta Region

Chapter 10
Analysis of the Factors Driving Wetland Degradation

Abstract Socioeconomic development has brought with it a large amount of immigration to the Yellow River delta region. Accordingly, the demand for food, accommodation, consumer goods, etc., has also been increasing. While on the one hand more manpower flowing in to the region helps development, on the other hand it means more land has to be transformed for farming, housing, and so on, and the urban area is also expanding. Under these conditions, the pressure on wetlands is increasing, and their coverage is shrinking. Over the past 20 years, petroleum exploitation and the oil processing industry have been booming. Inevitably, water, soil, and the ocean have suffered associated pollution from chemicals, organic matter, heavy metals, and so on, which jeopardize the flora and fauna living on wetlands. In view of these factors, the quality, health status, function, and positive effects of wetlands are in degradation. In recent times, influenced by climate change, annual rainfall has declined. Linked to this, water supply from the Yellow River has also reduced, with sometimes no water at all arriving in the delta area. Meanwhile, more and more freshwater is being exploited for human use, agricultural irrigation, and industry production. Hence, even more pressure is being placed upon water supply in the region and subsequently yet more strain on wetlands. Furthermore, in this region the evaporation–precipitation ratio is high (3.5:1), and the water table—as well as the content of soluble ions in the soil—is also high. Combined with its low altitude, these characteristics have resulted in the occurrence of secondary salinization. At present, there are 230,000 ha of salinized meadow soil and saline soil with a salt content of 0.4–2.0 % in the region. High soil salinity has adverse effects on plant growth and biodiversity, with few trees being able to grow there. As a consequence of the low forest cover and uneven distribution of vegetation, the existing ecosystem is weak and unstable.

Keywords Yellow River delta • Wetland • Restoration

J. Zhang, *Coastal Saline Soil Rehabilitation and Utilization Based on Forestry Approaches in China*, DOI 10.1007/978-3-642-39915-2_10,

As one of the three major ecosystems (i.e., forest, ocean, wetland), wetlands play an important role in coping with increasingly serious environmental problems. Apart from biological and water cleaning services, wetlands offer other functions that are beneficial to society, such as flood control, irrigation, drinking water reservoirs, and recreation (Brooks et al. 1991). In recent years, undisturbed natural wetlands, species-rich seminatural wetlands, and wetlands with a more or less intact hydrology have become very rare, and thus, there is increasing interest in the restoration of degraded wetlands (Brinson et al. 1981, 1998). Many of the problems concerning the eutrophication of aquatic systems are attributed to the intensification of agriculture, involving increased fertilization, the straightening of streams, and the lowering of groundwater tables (Ekstein and Hygnstrom 1996). As one possible solution to this problem, the construction, reconstruction, or restoration of wetlands has been initiated in many places worldwide. Indeed, the need for protection and restoration of wetlands has been recognized for a long time in environmental policy and nature conservation research (Euliss and Mushet 1996; Godfrey et al. 1985; Wakeley and Smith 2001). However, to ensure the success of such projects, information from different disciplines is urgently needed for both wetland management and the evaluation of the success of restoration and wetland construction principles.

The Yellow River delta (YRD) is formed by silts from the Yellow River and is hence very rich in natural resources. The crude oil, natural gas, chemical, salt, manufacturing, food processing, pulp, electricity, and bio-industries all benefit and are of primary importance in the area (Zhang and Li 2002). From the 1960s to 1990s, emphasis was on oil exploitation and other natural resources utilization, and as a result there was little emphasis on the protection of the natural environment. Consequently, the health status of the wetland worsened, and its ecosystem functions suffered. It is therefore highly significant to understand the causes of wetland degradation and instigate ecological restoration in the Yellow River delta region.

10.1 General Description of the Location

The Yellow River delta, located in Shandong Province (37°21′–38°12′N, 118°07′–119°10′E), is one of the biggest deltas in China with an area of 12,000 km^2. The area includes seven counties and 153,000 ha of national wetland reserve. The delta is located in the temperate zone and characterized by a monsoon climate.

Roughly 2,000 ha of newly formed land is created each year in this region by sand and soil sediment carried by the river. This soil is mainly composed of fine sand and is characteristically young with a high groundwater table, low fertility, and a tendency towards secondary salinization and desertification (Zhang 1997). The average groundwater table is generally 2–3 m deep, while along the coastline it is only 0.5–1.5 m. Due to encroachment of seawater and high evaporation, mineral concentrations in the soil are usually in the range 10–40 g/L and, in extreme cases, 200 g/L.

Fig. 10.1 The wetland landscape in the Yellow River delta region (Photograph taken by Zhang Jianfeng)

The delta is rich in biological resources, with an estimated 1917 species of flora and fauna, of which 50 are nationally protected and 47 are internationally protected. There are 393 plant species and varieties in the region, and most of them are salt tolerant. The natural vegetation is broadleaf deciduous forest (mainly hankow willow and weeping willow), shrubbery (mainly Chinese tamarisk), and shore coppice (see Fig. 10.1).

10.2 Wetland Types and Their Spatial Features

Generally, the Yellow River carries sands from the loess plateau at annual volume of around 10.5×10^8 t, of which one third ends up in the Bohai Sea and the remainder adds to the alluvial land of the delta region. On average, the area of newly formed land accumulated each year is approximately 21.3 km^2. While this land is muddy and wet, it nevertheless represents a clear trend of growing land formation in the delta region. It is estimated that there is a total wetland area of 333,427 ha

Table 10.1 Wetland types and their sizes in the YRD

Wetland type		Area (ha)	Percentage (%)
Natural	River and riparian	10,033	3.01
	Estuary	4,907	1.47
	Tidal land	8,425	2.53
	Worst salt-affected land	22,893	6.87
	Bulrush swamp	24,382	7.31
	Other marshland	17,602	5.28
	Woodland wetland	7,734	2.32
	Bush wetland	15,328	4.6
	Wet swamp	16,111	4.83
	Beaches	101,914	30.57
Artificial	Ditches	26,790	8.03
	Reservoirs	14,410	4.32
	Aquafarms	21,228	6.37
	Rice fields	19,103	5.73
	Ponds	18,846	5.65
	Salt fields	3,721	1.12
	Total	333,427	100

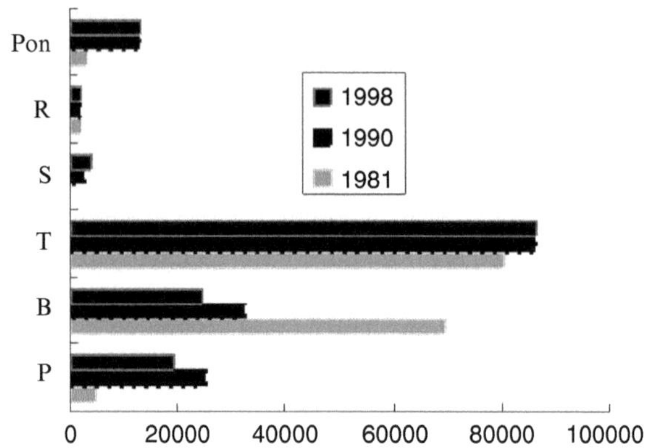

Fig. 10.2 Changes in some wetland types in the YRD region between 1981 and 1998: *P*, paddy fields; *B*, bulrush swamp; *T*, tidal land; *S*, salt fields; *R*, reservoirs; *P*, ponds

(42.7 % of the total delta area), of which natural wetland makes up 229,329 ha and artificial wetland 104,098 ha. The main types of wetland and their respective areas of coverage are shown in Table 10.1.

Influenced by many different factors, wetland structure and size is highly varied. The artificial wetland area in particular is expanding with social development. The areas of change for some wetland types are presented in Fig. 10.2.

It is clear that, between the early 1980s and the late 1990s, the areas of some of these wetland types have changed greatly, with a tendency for those wetland types that are useful for humans (e.g., ponds and paddy fields) to have grown.

10.3 Causes of Wetland Degradation

10.3.1 Insufficient Water Replenishment

In recent times, influenced by climate change, annual rainfall has declined. Linked with this, water supply from the Yellow River has also reduced, sometimes with no water at all arriving in the delta area. Meanwhile, more and more freshwater is being exploited for human use, for agricultural irrigation, as well as for industry. Hence, water shortages result in the wetland area decreasing (Gong 1986; Zhang 2004).

10.3.2 Serious Soil Salinization

The evaporation–precipitation ratio in the YRD region is high (3.5:1). Furthermore, the water table, as well as the content of soluble ions in the soil, is also high. Combined with its low altitude, the region has therefore suffered from the occurrence of secondary salinization. At present, there is around 230,000 ha of salinized meadow soil and saline soil with a salt content within the range 0.4–2.0 %. High soil salinity has adverse effects on plant growth and biodiversity (Zhang and Li 2002), with few trees being able to grow there. As a consequence of the low forest cover and uneven distribution of vegetation, the existing ecosystem is weak and unstable.

10.3.3 Irrational Land Use

Socioeconomic development has brought with it a large amount of immigration to the region. Accordingly, the demand for food, accommodation, consumer goods, etc., has also been increasing. While on the one hand more manpower flowing in to the region helps development, on the other hand much land has to be transferred to use for farming, housing, and so on. With the urban area also expanding, these conditions are causing wetland areas to shrink.

10.3.4 Pollution

Over the past 20 years, petroleum exploitation and the oil processing industry have been booming. Inevitably, water, soil, and the ocean have suffered from associated pollution from chemicals, organic matter, heavy metals, and so on. These pollutants jeopardize the flora and fauna living in the wetlands of the YRD by adversely affecting and degrading the quality, health status, function, and positive effects of the wetland ecosystem.

10.4 Ecological Restoration Techniques for Degraded Wetlands

10.4.1 Ensuring Water Supply

The YRD region is located in a semiarid monsoon zone, characterized by low precipitation with an uneven distribution (the amount of rainfall in summer makes up over 60 % of total annual precipitation), dry and windy conditions in spring, and very high soil evaporation in the season (Zhang and Li 2002), when the Yellow River is the major water source. In order to safeguard the health and functions of the wetland, it is necessary to supply additional water from the Yellow River or reservoirs. Additionally, other available biological measures can be employed, such as increasing vegetation coverage, decreasing soil evaporation, modifying soil structure, and enhancing moisture capacity. Whatever the measure or measures used, the retention of as much water as possible is highly important.

10.4.2 Protecting the Virgin Plant Community

There are some virgin plant communities growing in the wetland, such as *Nitraria* spp., *Tamarix* spp., *Aeluropus* spp., *Suaeda* spp., *Limonium* spp., *Salix matsudana*, *S. integra*, and *Phragmites australis*. Through natural selection and adaptation over millions of years, these plants have developed perfect structures to become tolerant to their adverse surroundings and grow well (Zhang et al. 2003a, b). However, increasing pressure from human activities, for example, over-pasturing, land exploitation, and oil pollution, the local flora has been damaged to a certain extent, leading to slower growth, reduced regeneration abilities, and declining vegetation cover. Thus, these plants must be protected, not only for conserving biodiversity but also for maintaining the health of the wetland.

10.4.3 Assisting the Regeneration of Virgin Flora

For the sake of preserving biodiversity, maximizing land coverage, and preventing the growth of local vegetation from being impaired, it is crucial to help virgin flora regenerate (Chen et al. 2002; Zhang et al. 2003c). As discussed above, under natural conditions, propagation of these plants can be difficult. However, a certain level of human assistance can be beneficial; for example, before the rainy season, digging ditches or holes close to trees for capturing more water and collecting more seeds has increased the coverage of *Tamarix chinensis* by 30 %. These measures can improve the living environment of species and help seeds to germinate and grow.

10.4.4 Regulating Land Use

As part of industrial development and urbanization, some parts of wetlands in the YRD region have been transferred to other land uses, often resulting in negative knock-on effects such as sewage problems, exhaust gas pollution, excess rubbish, and contamination from chemicals and fertilizers, all of which are harmful to the region's wetland areas. Hence, it is of vital importance to produce and maintain a comprehensive land-use plan aimed at removing these deleterious effects. Furthermore, treating pollution and restoring a clean and clear environment must also be part of the approach.

Unfortunately, owing to such adverse climate and soil conditions, the wetlands of the YRD region have been steadily degrading. Additionally, unfavorable human activities have also played a major role. To successfully protect and manage the wetlands of this region, long-term strategies are needed. It is necessary to develop guidelines for the management, design, evaluation, protection, conservation, delineation, and restoration of newly constructed, restored, and existing wetlands. Many of the approaches thus far encompassed by these developments are sectored, meaning they consider either biological parameters or abiotic features such as hydrology or nutrient dynamics. In facing these adverse conditions of the natural environment in the YRD region, ecological measures seem much more important and feasible. Interdisciplinary approaches are rare in spite of the many different functions of the wetlands, including habitat and regulation functions, as well as ecosystem services for human society. Previously, the focus has mostly been on wetlands as habitats for wildlife. However, during the last decade, the importance of wetlands as sinks and/or sources for nutrients and greenhouse gases has become increasingly prominent. Regulating land use is obligatory, especially in terms of reducing agricultural land use, as the presence of such land interrupts the natural path of water flowing into the wetland areas and also increases the potential for contamination. Introducing more plants is helpful not only for increasing vegetation coverage but also for tackling pollution by plant remediation, plant uptake, denitrification, and adsorption.

References

Brinson MM, Lugo AE, Brown S (1981) Primary productivity, decomposition, and consumer activity in freshwater wetlands. Annu Rev Ecol Syst 12:123–161

Brinson MM, Smith RD, Whigham DF et al (1998) Progress in development of the hydrogeomorphic approach for assessing the functioning of wetlands. In: Brinson MM (eds) Proceedings of INTECOL international wetland conference, Australian publisher, Perth

Brooks KN, Ffolliott PF, Gregerson HM et al (1991) Hydrology and the management of watersheds. Iowa State University Press, Ames

Chen SL, Jin KL, Fritz E et al (2002) Sodium and chloride distribution in roots and transport in three poplar genotypes under increasing NaCl stress. For Ecol Manage 168(1–3):217–230

Ekstein JD, Hygnstrom SE (1996) Fate of wetlands associated with the Central Nebraska irrigation canal system. Great Plains Res 6:41–60

Euliss NH, Mushet DM (1996) Water-level fluctuations in wetlands as a function of landscape condition in the prairie pothole region. Wetlands 16:587–593

Godfrey PJ, Kaynor ER, Pelczarski S et al (1985) Ecological considerations in wetlands treatment of municipal wastewaters. Van Nostrand Reinhold, New York

Gong HZ (1986) Silviculture in saline soil. China Forestry Press, Beijing

Wakeley JS, Smith RD (2001) Hydrogeomorphic approach to assessing wetland functions: guidelines for developing regional guidebooks; chapter 7, verifying, field testing, and validating assessment models, ERDC/EL TR-01-31, U S. Army Engineer Research and Development Center, Vicksburg

Zhang JF (1997) Review of saline soil utilization (in Chinese). J Shandong For Sci Technol 4:22–25

Zhang JF (2004) Agroforestry and its application in amelioration of saline soils in eastern China coastal region. For Study China 6(2):27–33

Zhang JF, Li XF (2002) Strategies to reclaim and ameliorate saline soil in the Yellow River delta region. In: Chi Chang, Brian Dobing (eds) International conference on environmentally sustainable agriculture for dry areas for the 3rd millennium proceedings, Dobing Enterprises, Canada, pp 264–269

Zhang JF, Li JY, Xing SJ (2003a) Experiment of seed germination under salinity stress. J Northeast For Univ 4:79–80

Zhang JF, Li JY, Xing SJ (2003b) Experiment on seed germination of *Chionanthus retusa* and *Cedrela sinensis* under salinity stress. J Beijing For Univ 4:88–90

Zhang JF, Li JY, Xing SJ (2003c) Advances of research on the mechanism of plant salinity tolerance and breeding of salt-tolerant plants. World For Res 16(2):16–22

Chapter 11
Impacts of Water Shortage

Abstract With an increasing population and rapid development of the economy and society of the Yellow River basin region, the Yellow River, the well known mother river of China, is at the edge of a survival crisis. The discrepancy between the supply and demand of water resources is a key issue. In 2000–2006, the mean annual discharge of the Yellow River entering into the delta was 13.2 billion m^3, a reduction of 18.6 billion m^3 compared with the 1980s, and 9 billion m^3 less than in the 1990s. The water requirement from various sectors is increasing. At present, the Yellow River runoff accounts for 2 % of the total in China; however, it is feeding 12 % of the population and represents 15 % of cultivated land in China. Seventy percent of water resources in the basin have been used, which is far beyond the internationally accepted warning level (40 %). The sustainable supplying capacity of the Yellow River water resources is facing a severe challenge. The most fragile ecosystem of the Loess Plateau has not been fundamentally changed. Facing such a severe outlook, sustainable use and management of the Yellow River is a priority.

Keywords Water shortage • Impact • Sustainable use • Yellow River Delta

The Yellow River is called "the cradle of Chinese civilization," as its basin is the birthplace of the northern Chinese civilizations and was the most prosperous region in early Chinese history. It originates in the Bayan Har Mountains in Qinghai Province in western China, flows through nine provinces of China, and empties into the Bohai Sea, where the Yellow River Delta is formed.

11.1 Basic Conditions of the Yellow River

The Yellow River basin has an east–west extent of 1,900 km and a north–south extent of 1,100 km. It means that the longest straight distance from east to west is 1,900 km, and the longest straight distance from north to south is 1,100 km. But

J. Zhang, *Coastal Saline Soil Rehabilitation and Utilization Based on Forestry Approaches in China*, DOI 10.1007/978-3-642-39915-2_11,

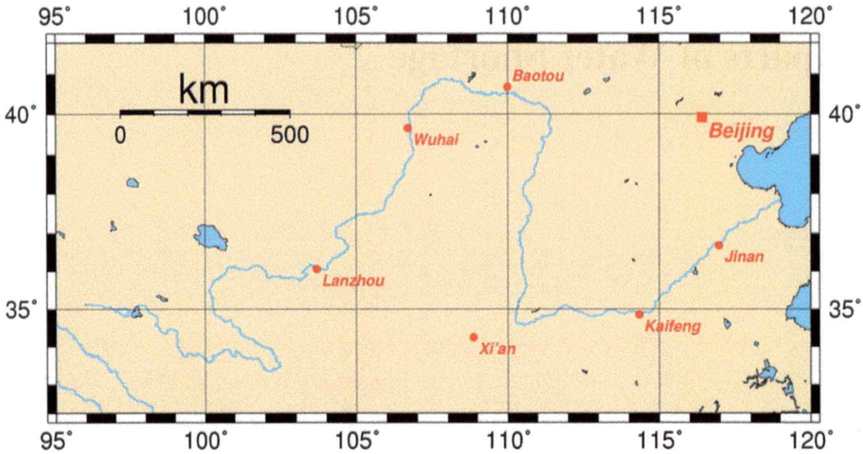

Fig. 11.1 Map of the Yellow River and the location of major cities along its course (Source adapted from Rolfmueller 2003)

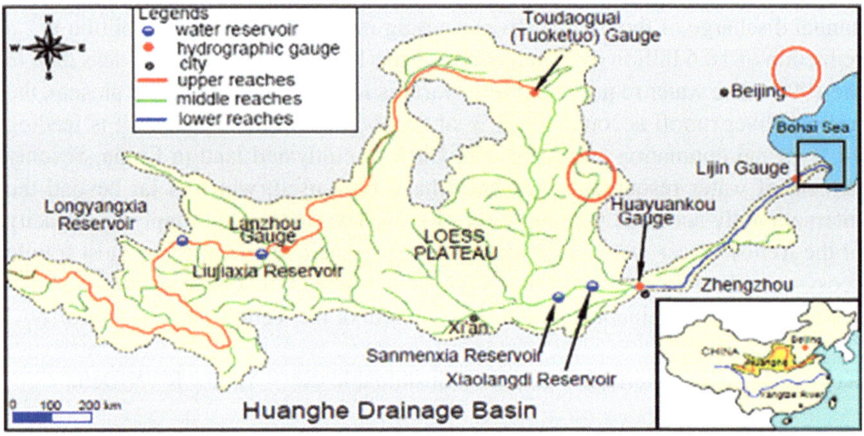

Fig. 11.2 Map of the spatial distribution pattern of the Yellow River system (Source adapted from Kevin 1987)

the river is not straight, in some place it turns curved, in this case the segment of the river becomes a "pond", whose length becomes longer. When we mention the total length of the river, usually we just calculate the straight distance, not the actual length of waterway (see Fig. 11.1).

The river is commonly divided into three stages: upper reaches, middle reaches, and lower reaches (see Fig. 11.2). The upper reaches of the Yellow River constitute a segment starting from its source in the Bayan Har Mountains and ending at Hekou County of Inner Mongolia just before it turns sharply to the north. This segment has a total length of 3,472 km and a total basin area of 386,000 km^2, 51.3 % of the total basin area. Along this length, the elevation of the Yellow River drops 3,496 m, with an average drop of 0.1 %.

The source section flows mainly through pastures, swamps, and knolls between the Bayan Har Mountains and the Anemaqen (Amne Machin) Mountains. The river water is clear and flows steadily. Crystal clear lakes are characteristic of this section. The two main lakes along this section are Lake Bob and Lake Eling, with capacities of 4.7 and 10.8 billion m^3, respectively. At elevations higher than 4,260 m above sea level, they are the largest two plateau freshwater lakes in China.

The part of the Yellow River between Hekou County in Inner Mongolia and Zhengzhou in Henan constitutes the middle reaches of the river. The middle reaches are 1,206 km long, with a basin area of 344,000 km^2, 45.7 % of the total, with a total elevation drop of 890 m, an average drop of 0.074 %. There are 30 large tributaries along the middle reaches, and the water flow is increased by 43.5 % in this stage. The middle reaches contribute 92 % of the river's silts.

The middle stream of the Yellow River passes through the Loess Plateau, where substantial erosion takes place. The large amount of mud and sand discharged into the river makes the Yellow River the most sediment-laden river in the world.

In the lower reaches, from Zhengzhou to the sea, a distance of 786 km, the river is confined to a levee-lined course as it flows to the northeast across the North China Plain before emptying into the Bohai Sea. The basin area in this stage is only 23,000 km^2, 3 % of the total. The total drop in elevation of the lower reaches is 93.6 m, with an average drop of 0.012 %.

The silts received from the middle reaches form sediments here, elevating the river bed. During 2,000 years of levee construction, excessive sediment deposits have raised the riverbed several meters above the surrounding ground. Few tributaries add to the flow in this stage; nearly all rivers to the south drain into the Huai River, whereas those to the north drain into the Hai River (Kevin 1987). The main tributaries of the Yellow River include the White River, Black River, Huang River, Zuli River, Qingshui River, Dahei River, Kuye River, Wuding River, Fen River, Wei River, Luo River, Qin River, and Dawen River.

The Yellow River Delta is located in the Yellow River Estuary in northeastern Shandong Province, China, and is situated between 117°31′–119°18′E and 36°55′–38°16′N. Currently, the total delta area is about 6,000 km^2. The delta was formed as a result of the continuous change in the flow path of the Yellow River on its way to the Bohai Sea. Because the Yellow River breached its banks in Tongwaxiang, in Henan Province in 1855, the original Xuhuai River flow path into the Bohai Sea is no longer active, and the river now flows into the sea through the Daqing River in northern Shangdong Province (Liu 2004; Zhang et al. 2010). The Yellow River Delta, where there is still a large area of land in a waste state because of soil salinity, is richly endowed with oil, gas, geothermal activity, and other natural resources including birdlife such as bitterns. The Shengli Oil Field located in the delta is the second largest in the country. Wetland and marine resources, a variety of biological species, and land resources make the territory an ideal place for sustainable development (Sun and Zhang 2006; Wei et al. 2008; Zhang 2005). The geographical location in the Bohai Economic Region, which is now a hot zone for development in China with good access to the Northeast Asia economic area, adds to the potential of the delta area.

11.2 Major Environmental Bottleneck

However, some major environmental problems exist in the area. In the past, owing to the low precipitation (560 mm annually), fresh water has mainly been obtained from the Yellow River. In recent years, as a result of the effects of climate change, there is less available water from the Yellow River because the water supply in the upper reaches of the river is diminishing. The limited supply of fresh water has caused a bottle-neck in further development in the region (Cao et al. 2006; Chinese Hydraulic Engineering Society, Yellow River Research Society 2003).

Globally, water shortage is a crucial issue and the gap between demand and supply is increasing (Braga and Granit 2003; Braga 2003; Graymore and Wallis 2010). In the Yellow River Delta, water has become a decisive factor restricting regional sustainable socioeconomic development (see Table 11.1). Therefore, to promote sustainable economic development and ecological protection, it is important to analyze the status of water resource development and rational use in the area, and to formulate countermeasures for the sustainable use of water resources (Geng et al. 2010; Guan et al. 2001).

The Yellow River plays a dominant role in forming and maintaining the regional hydrology. The river has its lowest discharge from late March to June. With increasing water consumption upstream, especially the rapid development of water diversion works for irrigation, the river flow frequently dries out. The main flood season is from July to October, when there is heavy rainfall in the middle and lower reaches. From October to mid-December, the river flow is steady. Then, from the end of December to March it will be ice bound, during which time the water level often rises drastically owing to poor drainage from ice blocking the river channel. The average flow rate is 1,330 $m^3\ s^{-1}$ with a maximum at flood peak of 10,400 $m^3\ s^{-1}$ and a minimum of 0 $m^3\ s^{-1}$. Water of the Yellow River has a pH of 8.0–8.3, a total hardness of 2.16–5.56 mg L^{-1}, and a mineral intensity of 0.2–0.6 g L^{-1}.

The annual increment of newly formed land from sediments carried by the river from 1974 to 1984 was 60.68 km^2, and from 1984 to 1994 was 23 km^2. The area of salt-affected soil is 4,430 km^2. These soils mainly comprise fine sands

Table 11.1 Statistics of population and land use in the Yellow River Delta (2006)

Population	1.839 million
Available land per capita (ha)	0.5
Arable land per capita (ha)	0.19
Percentage of arable land (%)	31.7
Percentage of forested land (%)	1.7
Percentage of grassland (%)	3.8
Inhabitable and mining land (%)	7.2
Road (%)	1.5
Water body (e.g., river, reservoir, pond, canal) (%)	30.9
Others (%)	23.2
Total (%)	100

from the Yellow River with characteristics of young age, a high ground-water table, low fertility, and a tendency to secondary salinization and desertification. The average ground-water table is 2–3 m; however, along the coastline it is only 0.5–1.5 m. Because of the encroachment of seawater and high evaporation, mineral concentration in the soil is usually 10–40 g L^{-1} and in extreme cases is 200 g L^{-1} (Sun and Zhang 2006). Population and land utilisation in the Yellow River Delta region are shown in Table 11.1 (Li 2006; Long et al. 2006).

11.3 Impact of Climate Change on Runoff in the Yellow River Delta Region

At present, changes in runoff in the Yellow River Delta closely relate to the inflow of water from the upper reaches, middle reaches, and major branches of the river (Liu 2007). Based on data of annual average rainfall and natural runoff depth in the drainage basin above Huayuankou, Henan Province, Liu reports that since the 1960s, the water cycle elements have shown a downward trend (Liu 2004). For the calculation of the annual runoff coefficient above Huayuankou, 1968 is regarded as the boundary to divide data from 1952 to 2001 into two stages, for which the runoff coefficients are 0.185 and 0.168 for the first and second stage, respectively. Liu reports that because of the impact of climate change and land-cover change, the runoff coefficient has decreased by around 9 %, resulting in a reduction in annual runoff of around 5.6 billion m^3.

The amount of flow from the Yellow River entering into the sea is calculated by using the runoff measured at the Lijin hydrological station, excluding industrial and agricultural water downstream of Lijin. Average annual flow entering into the sea in 1956–2000 was 31.32 billion m^3. In 1956–1979, the average annual flow was 40.98 billion m^3, whilst in 1980–2000 the average annual flow was 20.27 billion m^3 (nearly 50 % below that in 1956–1979). The amount of flow from the Yellow River entering into the sea shows a general downward trend, caused by the continuing reduction of natural water inflow as well as the ongoing increase in demand for water by the national economy (Hao et al. 2006).

The Yellow River Delta began to dry up in 1972, a phenomenon that has accelerated since the 1990s. From 1972 to 1999, periods of drying up in the lower reaches occurred 22 times among the 28 years, with dry periods occurring on average four times per 5 years, for a cumulative 1,092 days; 50 days a year. In 1997, the runoff measured at Huayuankou was the smallest amount since 1928, with a dry duration of 226 days over a river length of 704 km, creating a new historical record for the Yellow River (Guan et al. 2001; Zhang et al. 2005a).

The status of the ecological environment of the Yellow River Basin is closely related to the water resources. For example, wetland type and area change: it is estimated that the total wetland area is 3,334 km^2, this accounts for 42.7 % of the total delta area, of which 2,293 km^2 is natural wetland, and 1,041 km^2

is artificial wetland. Influenced by several factors, wetland structures and sizes vary. In particular, artificial wetland areas are expanding with social development (Zhang 2005). With the sharp drop in natural runoff, and the continued increase in water demand by rapid socioeconomic development in the Yellow River region, the disparity between supply and demand has become increasingly sharp.

Owing to a lack of correct understanding of the ecology of newly developed coastal areas, large-scale and irrational reclamation has been carried out since the end of the 1950s, e.g., the Bohai State Farm Bureau, established in 1959, cultivated a large area of "waste land" that resulted in the thorough clearing out of natural willows. Although reclamation may increase farmland area temporarily, secondary salinization sharply reduces productivity and often results in land abandonment. Kenli County possessed 59,000 ha of farmland in the mid-1950s, yet it had been reduced to 37,000 ha by 1980. This irrational cultivation has caused serious ecological deterioration in the Yellow River delta.

The oil industry is the main source of pollution in this region. It accounts for 40 % of the total waste gas, 40 % of the total waste water, and 43.9 % of the total industrial residues of Dongying. The main streams of the Yellow River and its large- and medium-sized reservoirs have fairly good water quality that meets the Class 3 National Surface Water Standard. However, the water quality of independent small rivers exceeds Class 5 of the same Standard because of the agricultural runoff and industrial sewage. Eutrophication and red tides have begun to occur in the river mouth, which are harmful to the aquatic ecosystem and decrease food production for waterfowls. In addition, poisonous pollutants affect birds throughout the food chain. With the rapid development of the local economy, pollution will inevitably be the main threat to this coastal ecosystem.

Meanwhile, as a substantial amount of water is diverted from the ecological environment, there has been a sharp drop in the amount of water flow entering into the sea (Fig. 11.3). Reduction of flow and drying up may result in serious channel shrinkage within the Yellow River Delta and rapid development of a secondary suspended river. This not only increases the possibility of embankments bursting or breaking, but leads to the shrinkage of wetlands, deterioration of water quality, and damage to biodiversity. According to statistics from the 1990s, as a result of the drying up of areas within the Yellow River Delta, the area of estuary vegetation decreased by almost half, fish numbers reduced by nearly 40 %, and bird numbers reduced by 30 % (Ma 2005; Shao et al. 2007). Therefore, a water shortage not only has an influence on the ecological environment but also has impacts on society and the economy. Since the start of the twenty-first century, water allocation and regulation of water-sediment measures have alleviated the environmental issues in the lower reaches and some estuary regions to an extent; however, functional drying up as well as ecological environment issues at the river basin level are far from being addressed.

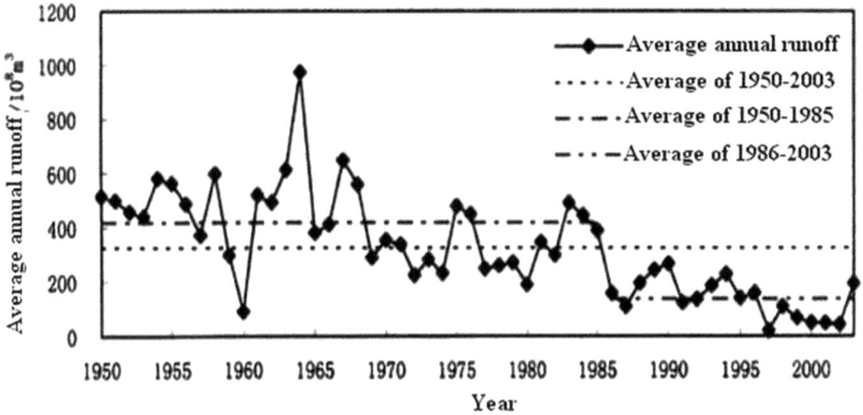

Fig. 11.3 Annual average runoff change at the Yellow River entrance (Source adapted from Mao et al. 2003)

11.4 Characteristics of the Water Resources

Water resources in the Yellow River Delta include the local water resources and the water transit. The local water resources are surface water and exploitable ground water (Mao et al. 2003). Restricted by the natural situation, water resources in the region are poor: surface runoff from rainfall (mean annual amount of 556 mm) is about 0.448 billion m^3; exploitable groundwater with a mineral concentration less than 2 g/L is very limited; most groundwater belongs to a saline, brackish, and brine classification; mineral content is very high; the rate of exploitation is very low; the shallow groundwater is about 0.043 billion m^3, and the annual exploitable groundwater is about 0.135 billion m^3.

The water transit is mainly comprised of the Yellow River, Xiaoqing River, and Zhimaigou River. The Xiaoqing River originates from Jinan in Shandong Province, and the mean annual discharge is approximately 0.528 billion m^3. Zhimaigou River originates from Zibo in Shandong Province, and the mean annual discharge is approximately 0.286 billion m^3. In recent years, a large amount of wastewater has been discharged into both of these rivers, resulting in serious water pollution, such that the water from both rivers cannot be used. In the past 10 years (1997–2006), the mean annual discharge of the Yellow River entering into the delta is about 11.2 billion m^3. This is the only exploitable fresh water in the Yellow River Delta region (Pan et al. 2005; Wang et al. 2004).

11.5 Current Statuses of Water Supply and Use

From 1990 to 2001 the mean annual water supply in the Yellow River Delta was 1.34 billion m^3. Of this amount, the Yellow River supplies around 1.25 billion m^3, accounting for 93 % of the total water supply, and is therefore the main water provider. Groundwater supply is about 0.91 billion m^3, mainly located in the south Xiaoqing River, accounting for 7 % of the total supply (Xie and Qi 2007; Zhang et al. 2009). The current major water-use sectors are agriculture, industry and oil production, urban services, and ecological requirements. The main water user is agriculture, followed by oil exploitation. The Shengli Oil Field uses 17 % of supplies. The largest discrepancy between water requirement and actual use is present in the ecological sector, which has the highest water requirement in the Yellow River Delta. According to an evaluation by Zhang (Zhang et al. 2005b), the minimum ecological water requirement is 1.921 billion m^3, and the maximum is 4.214 billion m^3. Therefore, it is evident that the water shortage is serious, especially for maintaining ecological balance.

11.6 Main Questions Relating to Water Exploitation and Sustainable Use

As previously mentioned, freshwater is scarce in the region, and water demand from oil development-oriented industry and comprehensive agriculture development is increasing. The Yellow River has played an indispensable role, and is a deciding factor in the economic development of the delta. However, the discharge of the Yellow River continues to decrease and cannot meet the water demand from human livelihood, economic production, and ecology. In addition, a large amount of water is wasted because of irregular use, and the water quality is deteriorating, leading to a more serious conflict between water supply and demand.

11.6.1 Discrepancy Between Water Supply and Demand

According to the data recorded from 1950 to 2006, the mean annual discharge from the Yellow River entering into the delta is about 31.8 billion m^3 (at the hydrological station). In the 1950s and 1960s, discharge was plentiful, and the amount of water entering into the sea was 48 and 50.1 billion m^3 during these periods, respectively. Since 1970, as a result of increased water use in the middle and upper reaches of the Yellow River, the amount of water entering into the delta has decreased. The mean annual amount of water entering into the delta was about 31.1 and 28.6 billion m^3 in the 1970s and 1980s, respectively. Compared with the mean annual Yellow River water entering into the delta, it is reduced by 0.67 and 3.2 billion m^3, respectively.

An estimated 14.1 billion m^3 of water entered into the delta in the 1990s, it is down by 17.7 billion m^3 and 14.5 billion m^3 since the 1970s and 1980s, respectively. The mean annual water entering into the delta from 2000 to 2006 is already 13.2 billion m^3, but it is not expected to exceed the amount of the 1990s (Xia et al. 2005; Xu 2004). "According to our observation that the mean annual water entering into the delta falls in recent years, lower than the amount of the 1990s."

The continuing decrease in the discharge of the Yellow River has severely affected the estuary. First, in industrial and agricultural production, some factories and enterprises have had to adjust production progress because of limited water availability, and some have closed or stopped production. Raw oil production has used saline water on occasion and has taken unusual measures such as closing some injection wells, resulting in serious loss of raw oil production. In agriculture, the growth of spring crops has been badly affected leading to a reduction in the grain harvest. Second, because of the build-up of sediment silt in the channel, medium-range floods have caused high water levels and flooding. This has resulted in severe flooding in the lower reaches of the Yellow River. Last, as a result of decreased flow into the delta, and almost no flow into the delta for much of the year except in wet periods, biodiversity has been threatened, resulting in degradation of the ecological environment (Wei et al. 2005; Xia and Zuo 2006).

11.6.2 Water Pollution

The water quality of the Yellow River Delta is deteriorating. Many rivers that have been polluted by industrial wastewater and sewage flow directly into the delta and into the Bohai Sea. The water quality of Xiaoqing River and Zhimai River exceeded level V of the standard for quality of surface water environment; water quality at Lijin station is between the III and IV levels. Water pollution threatens water supply safety, intensifies the conflict between water supply and demand, and results in direct harm on agriculture, fishery, livestock, and industry, seriously affecting people and future generations.

11.6.3 Water Wastage

Whilst there is water scarcity occurring in the delta, large amounts are also wasted. Because of low water prices in the delta region, industrial and agricultural water is wasted in large amounts. For instance, flooding channels are universally used in agricultural irrigation. However, the standard of supplementary facilities in the irrigation channel is very low, and the coefficient of effective water use is only 0.5. This is significant as irrigation quotas can reach as high as 3,000–3,750 m^3/ha. The reuse rate for industrial water is only about 20 %, and water consumption per output million for industry is 248 m^3. Fifty percent of water is lost between the urban water diversion site and the purifying station, and the rate of effective water use is only 55 %. These figures indicate that water wastage is a serious issue.

11.6.4 Over-Exploitation of Groundwater

The groundwater mineralization is very low in the very few areas suitable for well irrigation (Xu et al. 2005). In recent years, with the rapid development of industry and agriculture, water use has continued to increase and exploitation and use of groundwater has caused an imbalance between groundwater recharge and exploitation. As a result, the groundwater table has decreased, resulting in a funnel area of 200 km^2.

In addition, water regulation measures are currently inadequate. As the unified Yellow River water management and regulation is carried out only through administrative and technical measures, over-abstraction of water continues to occur in some places.

11.7 Countermeasures for Sustainable Use of Water

The pressure on the Yellow River is increasing, with the result that from the 1970s to the 1990s, the length of drying-up periods in the lower Yellow River increased. The entire basin ecosystem shows a trend of deterioration (Zhang 2006).

The carrying capacity of the river for nature and social development is limited. This should be maintained to enable the river to operate sustainably. Therefore, social and economic development plans have to consider the carrying capacity of the river as the first priority, and balance the water resource demand and supply accordingly.

11.7.1 Formulating a Comprehensive Governance Plan

The governance of the Yellow River estuary relates to many sectors and disciplines such as water conservancy, oil, agriculture, forest, livestock, fishery, communication, ocean, ecology, and the environment. At present, management of the Yellow River estuary lacks unified comprehensive planning, which is not helpful for sustainable economic development. Therefore, a review of the governance of the Yellow River estuary is urgently needed to clarify the direction and goals of governance, to speed up governance and development of the estuary, to provide the security to keep the Yellow River healthy, and to promote the development of an eco-society.

11.7.2 Reducing Water Wastage

With the development of the economy in the Yellow River Delta, the conflict between water supply and demand is becoming severe. Although there is serious water scarcity in the Yellow River Delta, the efficiency of water use is very low; large amounts of water are wasted by agriculture. To solve this conflict, the most effective way is to save water where possible and promote the development of a water-saving society. First, the layout of production and planning for economic development in the Yellow River Delta has to be based on the premise of water scarcity. We need to establish the concept that water demand is adjusted according to water supply. Dry-land agriculture must be developed and plant structure adjusted accordingly. The national industry policy must be strictly followed, and water assessment for construction projects enhanced and managed by a water use quota. Priority has already been given to treated wastewater, and the speed of increase in water demand has to be controlled. Next, the transfer of water rights must be constituted, making it possible to transfer agricultural water saving credits to industry. Currently, agricultural water-use accounts for 53 % of the total water in the delta. However, because the efficiency of agricultural water use is quite low, there is a large potential for reducing the water wastage. Water that has evaporated or infiltrated during irrigation can be saved and used for industrial projects. This could be via the implementation of agricultural water-saving projects by the owners of industrial projects who want to obtain water rights for their construction projects. The transfer of water rights will allow the market to fully play the leading role in water allocation. The economic measures of the water-saving mechanism need to be determined to improve efficiency for the rational development and the effective use of water resources. The goal of the development of a water-saving society will then be realized. Last, the rational mechanism of water pricing should be established to encourage the participation of the public and promote the socialization of water saving. Based on water scarcity and variation in water supply cost, if the Yellow River water prices are adjusted at the right time, the rational adjusting mechanism of water pricing is constituted. The adjustment of agricultural structure and water saving should be directed by the water price. The reasonable exploitation of groundwater should also be directed by the water price. We should encourage the participation of the public, and positively foster and develop water user associations that are involved in formulating water rights, water allocation, and water prices (Mao et al. 2003). The decision making, management, and supervision of water user associations should be carried out in a democratic manner to promote the social management of water saving.

11.7.3 Ensuring Basic Ecological Water Supply

Since the unified water regulation of the Yellow River in 1999, despite continuous drought, there has been no occurrence of drying up of the Yellow River in 8 consecutive years. The water demand for economic and social development within and outside the basin has been fully met. The amount of ecological water has effectively increased, with an improvement of the ecological environment (Wang et al. 2004). Therefore, we need to further enhance the unified management and regulation of the Yellow River for the future. For instance, a study on functional water for the regulation of the Yellow River should be performed. Based on the premise of keeping the Yellow River healthy, water regulation needs to consider economic water use, sediment transport water, ecological water use, and water-diluting pollutants in a holistic way. The total demand and the process demand of the above-mentioned four categories of water use have to be further clarified. The regulation of the ecological water demand in the estuary needs to be strengthened.

There must be strict control of the diversion of water between the Lijin station and the estuary. There is a distance of 100 km between the Lijin station and the estuary. There are ten water diversion projects within this section, and the total capacity of water abstraction is about 270 $m^3\ s^{-1}$. If the diversion of water is not controlled, there will not be enough water flowing into the wetlands. The amount of water that may be diverted should be controlled by the Yellow River water resources management, ensuring that the ecological water demand in the estuary is met.

11.7.4 Preventing Water Pollution

Water pollution prevention and control may be achieved by controlling total emission of pollutants and implementing the system of the pollutant emission license. With regard to water pollution control, considerable attention should be paid to the effect of the vegetation increase in the upper and middle reaches. In forested land, below the humus lies the tree roots, each clothed in fungal hyphae and the gels secreted by bacterial colonies. Thirty to forty percent of the bulk of the tree itself lies in the soil; most of this extends over many hectares, with thousands of kilometers of root hairs lying mat like in the upper 60 cm of soil (only 10–12 % of the root mass lies below this depth but the remaining roots penetrate as much as 40 m into the rocks below). The root mat actively absorbs the solution that water becomes, transporting it up the tree again to transpire to air. This is especially useful along waterways, especially in slop fields (Zhang et al. 2009). In this context, trees could be planned in combination with other practices to develop complete conservation systems that enhance landscape aesthetics, improve water quality, and provide wildlife habitats. Well-planned forest shelterbelts retain water and reduce evaporation by blocking drying winds in summer. Hence, it is significant to increase vegetation cover in the upper and middle reaches for the sustainable development in the lower reaches of the Yellow River Delta.

Moreover, we need to deal with water pollution issues in the Xiaoqing River and Zhimai River, and stringently control the emission of pollutants. Sources of point pollution, such as oil, paper, and chemical industries, must be strictly dealt with, the leakage of pollutants must be eliminated, and wastewater should be discharged following treatment only. Measures such as water saving and pollution control must be included in new projects and the standard on zero wastewater discharge has to be followed. The wastewater and garbage disposal infrastructure in cities must be developed. Garbage must be disposed of in a harmless way, and the rate of wastewater and garbage disposal should be increased. Fertilizers, pesticides, and plastic agricultural films should be used in a scientific and reasonable way to reduce non-point pollution caused by agriculture.

With an increasing population and rapid development of the economy and society of the Yellow River basin region, the Yellow River, the well-known mother river of China, is at the edge of a survival crisis. The discrepancy between the supply and demand of water resources is a key issue. In 2000–2006, the mean annual discharge of the Yellow River entering into the delta was 13.2 billion m^3, a reduction of 18.6 billion m^3 compared with the 1980s, and 9 billion m^3 less than in the 1990s. The water requirement from various sectors is increasing. At present, the Yellow River runoff accounts for 2 % of the total in China; however, it is feeding 12 % of the population and represents 15 % of cultivated land in China. Seventy percent of water resources in the basin have been used, which is far beyond the internationally accepted warning level (40 %). The sustainable supplying capacity of the Yellow River water resources is facing a severe challenge. The most fragile ecosystem of the Loess Plateau has not been fundamentally changed. Facing such a severe outlook, sustainable use and management of the Yellow River is a priority.

References

Braga BPF (2003) The role of regulatory agencies in multiple water use. Water Sci Technol 47(6):25–32, London, IWA Publishing

Braga B, Granit J (2003) Criteria for priorities between competing water interests in a catchment. Water Sci Technol 47(6):149–151, London, IWA Publishing

Cao WB, Wan L, Zeng YJ et al (2006) Impacts of global warming on the eco-environment in the headwater region of the Yellow River. Earth Sci Front 13(1):40–47

Chinese Hydraulic Engineering Society, Yellow River Research Society (2003) The symposium on the problems and countermeasures of the Yellow River Delta. Yellow River Conservancy Press, Zhengzhou

Geng Y, Mitchell B, Tsuyoshi F et al (2010) Perspectives on small watershed management in China: the case of Biliu. Int J Sustain Dev World Ecol 17(2):172–179

Graymore MLM, Wallis AM (2010) Water savings or water efficiency? Water-use attitudes and behaviour in rural and regional areas. Int J Sustain Dev World Ecol 17(1):84–93

Guan H, Zhang YH, Qin YP (2001) Analysis of the correlation among several cutting-out characteristic factors in the lower reaches of the Yellow River. J Hehai Univ (Nat Sci) 31(2):79–82

Hao ZC, Wang JH, Li L (2006) Impact of climate change on runoff in source region of Yellow River. J Glaciol Geocryol 28(1):1–6

Kevin S (1987) The Yellow River: a 5000 year journey through China. Child & Associates Publishing, Chatswood

Li B (2006) Preliminary analysis of the problem and regulation on the water resources ecology of the Yellow River Estuary. http://www.paper.edu.cn. Accessed 18 Aug 2006

Liu CM (2004) Research on water cycling evolution in the Yellow River basin. Adv Water Sci 15(5):608–614
Liu CZ (2007) The advances in studying detection of stream flow trend influenced by climate change. Adv Earth Sci 22(8):777–783
Long H, Mao LQ, Che ZX et al (2006) Impacts of climate change on Yellow River water resources. J Water Res Water Eng 17(4):74–77
Ma ZG (2005) History evolution rules and cause of formation of Yellow River runoff. Chin J Geophys 48(6):1270–1275
Mao HY, Zhao QJ, Gao Q (2003) New thought and pattern under the eco-environmental restriction in the Yellow River Delta. J Nat Resour 18(4):47–51
Pan ZQ, Liu GH, Zhou CH (2005) Temporal and spatial analysis of water demand of crops in Yellow River Delta based on remote sensing. Adv Water Sci 16(1):62–68
Rolfmueller (2003) Title of Huanghemap. http://en.wikipedia.org/wiki/File:Huanghemap.png. Accessed 28 Mar 2005
Shao XM, Xu YQ, Yan CR (2007) Spatial and temporal distribution analysis of water balance in the Yellow River basin. Clim Environ Res 12(1):74–80
Sun QX, Zhang JF (2006) Discussion on forestry sustainable development in the Yellow River Delta region. Chin Forest Sci Technol 5(4):63–67
Wang YZ, Zhang YL, Li R et al (2004) Analysis on evolution law and variation trend of droughts occurred in the middle Yellow River. Yellow River 26(8):43–45
Wei XH, Li WH, Zhou GY et al (2005) Forests and stream flow: consistence and complexity. J Nat Resour 20(5):761–770
Wei YC, Miao H, Ouyang ZY (2008) Environmental water requirements and sustainable water resource management in the Haihe River Basin of North China. Int J Sustain Dev World Ecol 15(2):113–121
Xia J, Zuo QT (2006) Advances of international hydrological science research. Adv Earth Sci 21(3):256–261
Xia J, Ye AZ, Wang GS (2005) A distributed time-variant gain model applied to Yellow River (I): model theories and structures. Eng J Wuhan Univ 38(6):10–15
Xie CW, Qi SH (2007) Analysis of climatic factors of runoff fluctuations in the cold area of the Yangtze and Yellow River Source area. Resour Environ Yangtze River Basin 16(2):251–255
Xu JX (2004) Runoff renewable change and its influencing factors in the reach of Hekouzhen-Longmen of the Yellow River. Adv Nat Sci 14(7):787–791
Xu YL, Huang XY, Zhang Y et al (2005) Statistical analyses of climate change scenarios over China in the 21st century. Adv Clim Change Res 1(2):80–83
Zhang JF (2005) Causes of wetland degradation and ecological restoration in Yellow River Delta region. Forest Study China 7(2):15–18
Zhang GH (2006) Analysis on potential effects of global climate change on natural runoff in the Yellow River Basin. Geogr Res 25(2):268–275
Zhang CC, Wang GQ, Wei JH (2005a) Study on water requirements in Yellow River Delta based on remote sensing. J Soil Water Conserv 19(1):149–152
Zhang XC, Liu CM, Li DY (2005b) Analysis on the water supply capacity of the Yellow River drainage basin. J Geogr Sci 60(1):79–86
Zhang JF, Jiang JM, Zhang ZJ (2009) Discussion on role of forest to control agricultural non-point source pollution in Taihu Lake basin based on source-sink analysis. J Water Resour Prot 1:345–350
Zhang JF, Chen GC, Xing SJ et al (2010) Carbon sequestration of black locust forests in the Yellow River Delta region, China. Int J Sustain Dev World Ecol 17(6):475–480

Part V
Case Study

Chapter 12
Establishment of *Tamarix chinensis* Stands and Its Amelioration of Saline Soil

Abstract *Tamarix* spp. is an important tree species for afforestation and plant community restoration in badly affected saline soils. In the present reported study, it was found that, after planting, this species can restrict salt returning to the soil surface and improve soil physical and chemical properties. Similar results have been found in other studies as well. It was concluded that the maximum salt concentration for seed germination of *Tamarix* is 0.7–0.8 %; for seedling survival it is about 1.64 %; and when the NaCl concentration reaches 1.0 %, cuttings are able to root normally. Salts have generally adverse effects on seed germination, and when salt concentrations rise, the germination rate will typically fall. The situation is similar for seedlings. Site preparation is the first step for forest establishment. In terms of environmental and soil traits in the Yellow River delta region, it was found that both ditch-shape and big-hole-shape site preparation are ideal for ensuring seedling survival. Furthermore, planting in spring or summer was found to be best, while direct seeding after rain is also a possibility.

Keywords *Tamarix chinensis* • Planting techniques • Salinity tolerance • Saline soil amclioration

The Yellow River generally carries on an annual basis around 10.5×10^8 t of sand from the loess plateau into the Yellow River delta region, of which one third ends up in the Bohai Sea and the remainder forms alluvial land in the delta area. On average, 21.3 km^2 of newly formed land is added each year, resulting in a total area at present of 8,100 km^2, of which 443,000 hm^2 is affected by salt. The soil area with salt content higher than 0.8 % is 233,000 hm^2. This soil is mainly composed of fine sand and is characteristically young with a high groundwater table, low fertility, and a tendency towards secondary salinization and desertification (Zhang 1997). The average groundwater table is generally 2–3 m, while along the coastline it is only 0.5–1.5 m. Due to encroachment of seawater and high evaporation, mineral

J. Zhang, *Coastal Saline Soil Rehabilitation and Utilization Based on Forestry Approaches in China*, DOI 10.1007/978-3-642-39915-2_12,

concentrations in the soil are usually 10–40 g/L and, in extreme cases, 200 g/L. Most of these soils are uncultivated due to salinity. Furthermore, some of the land that has been forested is degrading because of improper tree species selection or unsuitable planting techniques. In order to restrict soil salinization and improve land quality, planting salt-tolerant trees is important.

12.1 Introduction

Tamarix spp. is characterized by its favoring of light and its tolerance of wet soil, salt, wind, and sand, as well as mead soils (Bohnert et al. 1995; Zhang 2003). In the past, it was the dominant species in the plant community of the Yellow River delta (YRD) region. However, owing to human population growth and inappropriate management practices, its dominance has declined significantly (Zhang and Qin 2007). As a result, the area of land in this coastal region suffering from secondary salinization is expanding, and salt concentrations in soils are rising (Zhang and Li 2002). Given the problems the area faces in this regard, it has been identified as a good case to test the success of afforestation in combating such adverse coastal environmental conditions (Qureshi and Barrett-Lennard 1998). Accordingly, to protect the environment from further degradation and to restore the local plant community as much as possible, salt tolerance selection and planting techniques for *Tamarix chinensis* were tested.

Test plots were located in the Yellow River delta region (37°21′–38°12′N, 118°07′–119°10′E), which is an area that includes seven counties and a 153,000-ha national wetland reserve. It lies in the warm temperate zone and is characterized by monsoon climate, with an annual mean temperature (T) of 11.4 °C, maximum T of 40 °C, minimum T of −19.3 °C, annual daylight hours of 2,305 h, annual mean rainfall of 556.8 mm, and annual evaporation of 1,948 mm (Zhang 2002). The soil is mainly salt-affected meadow and coastal solonchak. Vegetation coverage in the test fields mainly comprised halophytes, such as *Tamarix* spp. (see Fig. 12.1), *Suaeda salsa*, *Limonium sinense*, and *Phragmites* spp. These lands are hindered by adverse climatic, soil, and environmental factors, for example, drought, waterlogging, salinity, wind, and pollution. Thus, soil quality and productivity in the region are not conducive to economic development, and the level of soil degradation affects the ecological balance as well as environmental stability (Zhang 2004).

12.2 Materials and Methods

The experimental materials were collected from the local area. Seeds of *Tamarix chinensis* were harvested from the counties of Hekou and Zhanhua throughout the year, and cuttings used were one-year-old. Seed germination under salinity stress was carried out through a combination of field tests and lab experiments, with six

Fig. 12.1 The growing of *Tamarix chinensis* in the YRD region (Photograph taken by Zhang Jianfeng)

Table 12.1 Site preparation details

Site no	Type	Size (length × width × depth) (cm)
1	Hole shape	30 × 30 × 30
2	Hole shape	40 × 40 × 40
3	Hole shape	50 × 50 × 50
4	Hole shape	100 × 50 × 40
5	Ditch shape	(500–1,200) × 50 × (30–35)
6	Ditch shape	(500–1,200) × 50 × (15–20)
7	Ditch shape	(300–800) × 30 × (15–20)
8	Ditch shape	(300–800) × 30 × (10–15)

salt treatments and pure water as the control experiment. The seedling salt tolerance experiments were conducted in the field, while cutting salt tolerance was tested using four different NaCl concentrations, with each treatment applied to 20 cuttings. To test different planting techniques, eight types of site preparation were examined, as detailed in Table 12.1.

The effect of *Tamarix* on soil improvement was tested by selecting 5–10-year-old *Tamarix* forests, collecting soil samples at different soil depths, and analyzing the soil salt content and other physical properties, with bare land used as the control.

12.3 Results and Analysis

Tamarix spp. extracts salt in a typical way, that is, it absorbs salt from the soil through its roots and extracts it from leaves. Moreover, its tolerance to salt is not only a feature of its leaves but is also present in seeds, seedlings, and cuttings.

Table 12.2 Effect of salt on seed germination in *Tamarix* spp.

NaCl (%)	0	0.4	0.6	0.8	1.0	1.2	1.5
Germination rate (%)	91.3	87.4	69.2	20.2	7.8	1.3	0.3

Table 12.3 Effect of NaCl on rooting of *Tamarix* cuttings

NaCl (%)	0	0.3	0.5	1.0	3.0
Rooting rate (%)	96.5	95.1	92.5	80.4	0

The seed germination experiments conducted in the field indicated that *Tamarix* can grow well in soil with NaCl concentrations of 0.63 % and 0.721 % after germination. The lab-based experiment on seed germination (see Table 12.2) indicated that seeds could germinate as usual when salt content is 0.4 %, but when the NaCl concentration reached 0.6 %, the germination rate fell to 70 %, and when the NaCl concentration was increased to 0.8 %, the germination rate declined obviously. Considering the results of both the lab- and field-based experiments, it can be concluded that the maximum NaCl concentration for seed germination is 0.7–0.8 %.

Planting time was found to have a significant effect on the survival rate of *Tamarix*. When planting in November, the survival rate was 23.3 %, while planting in March increased the rate to 50.6 %, and planting in August raised it further still to 82.6 %. The cause was perhaps that groundwater in winter rose via soil capillary action due to drought and, then, saltwater was returned to the soil surface from the deep profile. Consequently, the soil salt content increased and was harmful for the survival and growth of seedlings. However, in August, which is a very wet time of the year in the region, soil salts would exit the topsoil and pass into deeper parts of the profile via infiltration, thus alleviating the adverse effects on the seedlings.

The results of the salt tolerance experiments indicated that when the salt concentration is 1.64 %, nearly all *Tamarix* seedlings will likely die. Whereas, if the salt concentration is no higher than 1.203 %, seedlings will grow normally, and if it is below 1.0 %, seedlings will grow well.

Cuttings represent important propagation material for *Tamarix* (Zhang et al. 2004). Their rooting rate results under salt stress are given in Table 12.3, in which it is shown that when salt concentration is below 0.5 %, *Tamarix* cuttings can root normally, but if the salt concentration increases to 1.0 %, the rooting rate falls to 80.4 %, and if the concentration is as high as 3.0 %, no rooting at all occurs. Thus, the optimum soil salt concentration for *Tamarix* cutting survival is below 1.0 %.

Site preparation is key to successful afforestation by direct seeding. Effective site preparation can decrease soil salt content, as well as control weeds. In the present reported study, different types of site preparation had significant effects on germination and growth of *Tamarix*. The relationships among them are shown through the information presented in Tables 12.4, 12.5, 12.6 and 12.7.

From Table 12.5, it can be seen that soil salt concentration declined after all types of site preparation, but was best for type 5 (i.e., site size of (5–12) × 50 × (30–50)

Table 12.4 Changes in *Tamarix* survival rate when planting in different seasons

Test plot	Site condition[a]	Site preparation	Surviving rate (%)
1	I	Big hole	93.0
2	III	Big hole	85.0
3	I	Ditch	21.0
4	III	Hole	40.4
5	II	Hole	11.9
6	I	Hole	20.9
7	I	Hole	57.4
8	I	Hole	69.3
9	III	Big hole	50.0
10	II	Hole	22.5

[a]I rather good, II rather bad, III very bad

Table 12.5 Variation in soil salt content following different types of site preparation

Site preparation type	1	2	3	4	5	6	7	8
Salt concentration before site preparation (%)	2.579	2.579	2.579	2.579	2.579	2.579	2.579	2.579
Salt concentration after site preparation (%)	1.612	1.362	0.983	1.172	0.861	1.198	1.64	2.429
Salt reduction (%)	0.967	1.217	1.569	1.407	1.718	1.381	0.939	0.15
Order by level of salt reduction	6	5	2	3	1	4	7	8

Table 12.6 Effect of site preparation on germination (stock per unit)

Site preparation type	1	2	3	4	5	6	7	8
Block I	76	104	102	101	111	83	159	128
Block II	56	126	86	149	106	113	112	96
Block III	127	108	107	96	203	94	25	104
Mean	86.3	112.7	98.3	115.3	140	96.7	113	109.3

Table 12.7 Growth of seedlings following different types of site preparation (cm)

Site preparation type	1	2	3	4	5	6	7	8
Block I	22.3	36.3	40.4	33.8	47.4	24.9	28.2	22.2
Block II	33.5	33.9	51.1	47.0	63.0	37.2	25.8	28.0
Block III	30.2	37.8	52.7	55.8	53.4	32.0	29.0	23.0
Mean	28.7	36.0	48.1	45.5	54.6	31.9	27.3	24.4
Mean of dominant trees	48.9	51.2	79.5	66.5	75.3	52.9	51.0	31.7

cm), and then types 3 and 4 (i.e., site sizes of $50 \times 50 \times 50$ cm $100 \times 50 \times 40$ cm, respectively). The worst was type 8 (i.e., $(300–800) \times 30 \times (10–15)$ cm). Additionally, from Table 12.6 it seems that ditch-shaped site preparation was better than hole-shaped, and the results in Table 12.7 also show similar results. To summarize, to promote good growth, the order in terms of the best site preparation type was

Table 12.8 Comparison of the physical properties of *Tamarix* woodland soil and bare land

Site	Soil depth (cm)	Soil moisture (%)	Total porosity (%)	Soil capacity (g/cm^3)	Organic matter (%)	Total salt (%)	Mean (%)
Bare land	0–20	20.12	45.18	1.45	0.2459	1.78	1.034
	20–40	28.39	46.42	1.42	0.1112	1.00	
	40–60	21.63	46.79	1.41	0.0448	0.71	
	60–100				0.0785	0.84	
Tamarix woodland	0–20	31.60	53.91	1.20	0.3250	0.59	
	20–40	28.58	47.92	1.38	0.1336	0.48	
	40–60	29.08	44.53	1.45	0.0336	0.53	0.544
	60–100				0.1009	0.56	

type 5, type 3, and type 4. The worst site preparation type was type 8. This means that ditch-shaped site preparation (with the length of the ditch being above 5 m, the width being 50 cm, and the depth over 35 cm), as well as big-hole-type preparation, is best suited for direct seeding of *Tamarix* in badly affected saline soil.

Tamarix forest could also improve soil quality. To examine this, one woodland of age 5–10 years was surveyed in November, and the physical properties of the soil were analyzed (see Table 12.8).

As can be seen from the results in Table 12.8, *Tamarix* forest effectively reduces the soil salt concentration, especially at depths of 0–40 cm in the soil profile. Furthermore, soil organic matter, soil moisture, soil capacity, and soil total porosity are also positively affected.

12.4 Conclusions and Discussion

Tamarix spp. is an important tree species for afforestation and plant community restoration in badly affected saline soils. In the present reported study, it was found that, after planting, it can restrict salt returning to the soil surface and improve soil physical and chemical properties. Similar results have also been reported in other experiments (Gong 1986).

It can be concluded from the present reported experimental results that the maximum salt concentration for seed germination of *Tamarix* is 0.7–0.8 %; for seedling survival it is about 1.64 %; and for cuttings, when the NaCl concentration reaches 1.0 %, they are able to root normally. Salts generally have an adverse effect on seed germination, and as the salt concentration in soil rises, germination rates usually fall (Khan and Gul 1998; Pujol et al. 2000). The situation is similar for seedlings (Bandyopadhyay 1988; Chen et al. 2002; Crouch and Honeyman 1986).

Site preparation is the first step in forest establishment (Zhang 2002). From the results of the present study, in terms of environmental and soil traits in the YRD region, both ditch-shape and big-hole-shape site preparation was found to be ideal for ensuring seedlings survival. Planting in spring or summer was found to be better, while direct seeding after rain is also a possibility.

References

Bandyopadhyay AK (1988) A note on the effect of salinity on the growth of *Eucalyptus* and *Casuarina* seedlings. Indian For 114(6):343–344

Bohnert HJ, Nelson DE, Jensen RG (1995) Adaptation to environmental stresses. Plant Cell 7:1099–1111

Chen SL, Li JK, Fritz E (2002) Sodium and chloride distribution in roots and transport in three poplar genotypes under increasing NaCl stress. For Ecol Manage 168(1–3):217–230

Crouch RJ, Honeyman MN (1986) The relative salt-tolerance of willow cuttings. J Soil Conserv 42(2):103–104

Gong HZ (1986) Silviculture in saline soil. China Forestry Press, Beijing

Khan MA, Gul B (1998) High salt tolerance in germinating dimorphic seeds of *Arthrocnemum indicum*. Int J Plant Sci 159(5):826–832

Pujol JA, Calvo JF, Diaz LR (2000) Recovery of germination from different osmotic conditions by four halophytes from Southeastern Spain. Ann Bot 85:279–286

Qureshi RH, Barrett-Lennard EG (1998) Saline agriculture for irrigated land in Pakistan: a handbook. Australian Centre for International Agricultural Research, Canberra

Zhang JF (1997) Review of saline soil utilization. J Shandong For Sci Technol 4:22–25

Zhang JF (2002) Research on saline soil amelioration and forestation techniques. J Northeast For Univ 30(6):124–129

Zhang JF (2003) Advances of research on the mechanism of plant salinity tolerance and breeding of salt-tolerant plants. World For Res 16(2):16–22

Zhang JF (2004) Agroforestry and its application in amelioration of saline soils in eastern China coastal region. For Study China 6(2):27–33

Zhang JF, Li XF (2002) Strategies to reclaim and ameliorate saline soil in the Yellow River Delta region. In: Chi Chang, Brian Dobing (eds) International conference on environmentally sustainable agriculture for dry areas for the 3rd millennium proceedings. Dobing Enterprises, Canada, pp 264–269

Zhang JF, Qin GH (2007) Poplar-based agro-forestry in China and its economic analysis. Shandong Science and Technology Press, Jinan

Zhang JF, Xing SJ, Zhang XD (2004) Principles and practice of forestation in saline soil in China. Chinese For Sci Technol 3(2):62–70

Chapter 13
Effects of *Nitraria sibirica* on the Improvement of Salt-Affected Soil

Abstract *Nitraria sibirica* is a salt-tolerant plant, especially the provenance from Dayilu. The present reported experimental study reveals that this species is very good at preventing salt from accumulating as well as improving soil physical properties and enhancing soil nutrition. When a forest is going to be established, site choice and the site preparation method are very important. Related to this, because *Nitraria* seedlings are weak at the very beginning, their capacity to compete with weeds is rather low, and thus site preparation using methods such as continuous plowing and harrowing is necessary to wipe out these competing weeds. When it comes to planting *Nitraria* seedlings, container seedlings are better than bare-root seedlings, and planting in the rainy season is better in practice because of the local climate features. Through the experiments described in this chapter, it is possible to conclude that *Nitraria sibirica* has good salt resistance, wide development potential, and high ecological and economic benefits; hence, it could be applied in vast areas of serious salt-affected coastal soil. In terms of seed provenances, Dayilu was found to be an excellent source of *Nitraria sibirica*. When planting *Nitraria sibirica*, the site should be selected in salt-affected areas with salt content lower than 0.6–0.8 %; the site preparation method of two plowing two harrowing should be used; a planting density of 2505/ha is best; and container seedlings planted in the rainy season, or bare-root seedlings for use in direct planting for forestation, are the best strategies.

Keywords *Nitraria sibirica* • Planting techniques • Salt-affected soil

The Yellow River delta (YRD) region covers a total area of 8,100 km^2, of which salt-affected soil makes up 443,000 ha and the soil area with a salt content of greater than 0.8 % is 233,000 ha. Most of these soils remain uncultivated because of their saline conditions. Furthermore, some land that has been planted with trees is degrading because of poor tree species selection or unsuitable planting techniques having been applied (Zhang 2004).

J. Zhang, *Coastal Saline Soil Rehabilitation and Utilization Based on Forestry Approaches in China*, DOI 10.1007/978-3-642-39915-2_13,

13.1 Introduction

Nitraria sibirica belongs to the family *Zygophyllaceae*, possesses a well-developed root system, favors light, and is frost and salt tolerant. Furthermore, it has adapted to grow in sandy and/or saline soil and could improve soil properties and prevent soil erosion from wind (Niknam and McComb 2000). Various kinds of trace elements and amino acids have been detected in the pulp, core, and fresh juice of *Nitraria sibirica*. Moreover, the biological pigments that are richly present in the juice of its fruit, and the rind, are stable under conditions of light, heat, and oxygen (Ashraf 1994; Gupta 1987).

In Shandong Province, *Nitraria sibirica* grows wildly in areas along the coast of the Bohai Gulf and Laizhou Gulf (see Fig. 13.1). In order to restore the plant community as soon as possible in heavy salt-affected soil in the YRD region, *N. sibirica* was chosen as one of the important target plant materials, and salt tolerance experiments and planting techniques have been studied since 1994.

Test plots were created in the YRD region, which is situated in the warm temperate zone characterized by a monsoon climate: annual mean temperature (T) of 11.4 °C, maximum T of 40 °C, minimum T of −19.3 °C, annual daylight hours of 2,305 h, annual mean rainfall of 556.8 mm, and annual evaporation of 1,948 mm. The soil is mainly saline meadow and coastal solonchak. The flora in the test field is mainly halophytes, such as *Tamarix* spp., *Suaeda salsa*, *Limonium sinense*, and *Phragmites* spp.

Fig. 13.1 The growing of *Nitraria sibirica* in the YRD region (Photograph taken by Xing Shangjun)

13.2 Materials and Methods

Seeds of *Nitraria sibirica* came from Dayilu in Hekou District, Dongba in Liuhu Town, Heidao in Zhanhua County, Yangzhuang in Shouguang City, and Tanyang Forest Farm in Wudi County.

Fixed plots combined with typical plot investigations were used to perform the salt tolerance experiments. Fixed plots were established at No. 4 Farm and the Forest Research Institute in Dongying City, and six treatments according to the salt content in the soil (depth of 0–25 cm) with three duplications were employed. All seedlings were container seedlings.

There were four types of planting techniques: (i) planting trees following site preparation; (ii) site preparation in early spring and planting during the rainy season; (iii) consecutive (twice) plowing and harrowing, that is, site preparation lasting 2 years, with all plowing in early spring, first harrowing in summer to wipe out weeds and then second harrowing before tree planting; and (iv) hole-shaped site preparation at a size of 30 × 30 × 25 cm, with planting during the rainy season.

Forest planting types were direct seeding, container seedling planting, and bare-root seedling planting (Gupta et al. 1987). Soil samples were collected in order to analyze soil nutrition, salt content, and soil physical properties 3–4 years after planting (Tabatabai 1994; Malcolm and Sumner 1998). Samples were taken from 1.5 m inside and outside the crown area, with bare land as the control. Survival rate and living stock investigations were respectively carried out 1 month and 1 year after planting (Chen et al. 2002; Gulzar et al. 2001).

13.3 Results and Analysis

It can be seen from Fig. 13.2 that the seedlings of *Nitraria* are capable of salt resistance.

When surface soil salt content was 0.6 %, almost all seedlings survived. However, when salt content rose to 0.8 %, the survival rate fell to 80 %, and when the salt content rose further still to above 0.8 %, the survival rate fell markedly. When the salt content rose to 1 %, the survival rate was just 50 %.

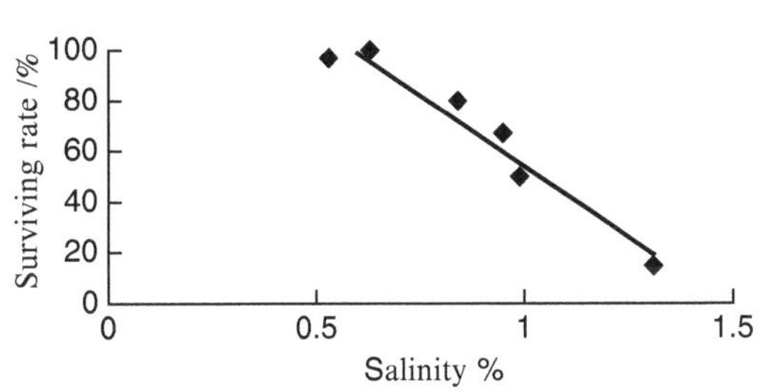

Fig. 13.2 Survival rate of *Nitraria* under salt stress

Table 13.1 The effect of soil salt content on the growth of *Nitraria*

Planting date	Salt content (%)	Growth performance	Shoot length (m)	Crown (m^2)
Aug. 1995	2.57	+	0.2–0.5	0.28
	1.74	++	0.8–1.0	1.20
	1.02	+		2.90
	0.58	++		10
Jun. 1994	1.32	++		3.20
	0.91	++		6.33
	0.29	+		3.80

++ very good; + good

Table 13.2 Effect of soil salt on the growth of *Nitraria*[a]

Salt content (%)	Crown (m^2)	Number of branches	Length of branch (m)	Diameter of branch (cm)
0.8–1.0	0.098	4	0.19	<0.2
0.6–0.8	0.53	3	0.43	0.23
<0.6	0.04	4	0.13	<0.2
1.0–1.2	2.19	9	0.69	0.33
0.8–1.0	4.10	10	1.01	0.56
<0.6	0.135	4	0.24	0.21
1.0–1.2	4.31	9	1.10	0.76
0.8–1.0	4.69	13	1.17	0.71
1.0–1.2	6.37	10	1.33	0.97
0.8–1.6	7.80	15	1.23	0.82
0.8–1.1	6.23	13	1.22	0.79
<0.6	4.14	13	0.91	0.76

[a]Planting date was 10 June 1994

The investigation showed that, when salt content is below 0.6 %, one can expect the survival rate of *Nitraria* seedlings to be 92–96 %, but if it is higher at 0.6–0.8 % or 0.8–1.0 %, one can expect the survival rate to fall to 80–88 % and 50–70 %, respectively. Meanwhile, if the salt content exceeds 1 %, one should expect a survival rate of lower than 4 %. These findings indicate that the maximum soil salt content is 0.8 % to ensure strong survival rates when planting *Nitraria*. However, it was found that, 2 years after planting, *Nitraria* still grew well when soil salt content was 1.5–2.0 %. Furthermore, after 3 years, salt could not influence its growth (see Tables 13.1 and 13.2).

Results also showed that different provenances of *Nitraria* had different abilities in terms of their salt resistance. Among five provenances, Dayilu was the best. When the soil salt content reached 1.74 %, survival rate still remained at 94 %, and when 2.57 %, it was still above 90 %, and trees grew well. The other four provenances showed no differences (see Table 13.3).

Tests designed to examine the influence of site preparation showed that different methods could effectively decrease the soil salt content. Furthermore, such methods

Table 13.3 Effect of provenance on the growth of *Nitraria*

Provenance	Salt content (%)	Survival rate (%)	Growth performance	Shoot length (cm)
Dayilu	1.74	94	++	80~100
The other	2.57	91	+	20~50
Four provenances	1.71	3	–	
	1.31	15	–	20~30
	1.10	51	+	30~40
	0.84	81	++	80~100

++ very good; + good; ––very bad; –bad

were also effective at combating weeds, which benefited seedling establishment through reduced competition for nutrients. Jug et al. (1999) and Kovda (1983) also found similar results from their experiments. The influence of site preparation types on survival rate, living stock, and growth of *Nitraria* was significant. Among the four types, the effect of No. 3 (i.e., two-plowing-two-harrowing) was the optimum. After site preparation, the soil salt content decreased from 1.5–2.0 % to 0.8 %, and the survival rate of seedlings reached 96 %. Moreover, owing to weeds dying out, trees later grew very well. Method No. 2 was able to prevent evaporation and restrict salt returning to the soil surface owing to plowing in early spring, and thus it also had a certain effect on salt reduction, for example, from 0.6–1.5 % to 0.4–1.0 %, the survival rate of seedlings remained at 92 %. The other two types of site preparation were not ideal.

For the planting of forests, direct seeding was found not to be practical because the seeds of *Nitraria* are very small; seed germination in salt-affected soil was difficult. Additionally, weeds had a negative effect on growth. However, planting bare-root seedlings was viable; their survival rate reached 89 % in sites with a salt content of 0.6–1.0 % and grew well. An advantage of this approach is the lower cost, and therefore, it could be put into practice when salt content is below 1 %. The planting of container seedlings is more commonly employed, and their survival rate maintained above 90 % when salt content was within the range 0.6–1.5 %. The cause was that the effect of salt on growth was reduced to its lowest level. However, its cost was a little higher of course.

Soil salinity was found to have an influence not only on survival rate and shoot growth but also on crown growth. Figure 13.3 clearly illustrates this by showing that, as soil salinity rose, the crown of the tree grew more slowly. However, compared with Fig. 13.2, it can be seen the effects of salinity on survival rate and crown growth were different. The effect on survival rate was almost linear, while the effect on crown growth was more curved, which possibly suggests the latter involves more factors.

Table 13.4 indicates that *Nitraria* can effectively restrict the accumulation of salt in soil, especially at the surface. Also apparent is that the soil physical properties and soil nutrition in woodland were quite a lot better than on bare land. Thus, it can be concluded that *Nitraria* is suited to and effective at improving soil quality, as well as increasing land productivity.

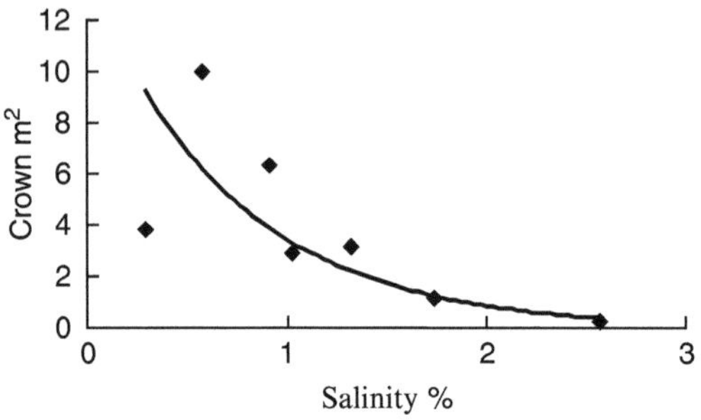

Fig. 13.3 Effect of soil salt content on crown growth of *N. sibirica*

Table 13.4 Soil physical properties at different site locations

Site	Soil depth (cm)	Soil moisture (%)	Volume weight (g/cm^3)	Soil porosity (%)
Woodland 1	0–20	24.92	1.17	55.85
	20–40	23.38	1.25	52.83
	40–60	24.56	1.35	49.06
	0–20	23.09	1.27	52.08
Woodland 2	20–40	26.99	1.28	51.70
	40–60	26.74	1.33	49.81
Control (bare land)	0–20	27.30	1.46	44.91
	20–40	29.57	1.38	47.92
	40–60	29.59	1.44	45.66

13.4 Discussion and Conclusion

Nitraria is a salt-tolerant plant species, especially the provenance from Dayilu. It can restrict salt accumulation, as well as improve soil physical properties and enhance soil nutrition (Zhang 2003). When a forest is going to be established, site choice and the site preparation method are very important. Because *Nitraria* seedlings are weak at the very beginning, and their capacity to compete with weeds is rather low, site preparation techniques such as continuous plowing and harrowing is necessary to wipe out the weeds. Container seedlings are better than bare-root seedlings, and planting in the rainy season is more practical in view of the local climate features. Overall, it can be concluded from the study described in this chapter that *Nitraria sibirica* has good salt resistance, wide development potential, and high ecological and economic benefits and could be applied across vast areas of salt-affected coastal soil. In terms of seed provenances, Dayilu is an excellent seed source of *Nitraria sibirica*. Furthermore, when planting *Nitraria sibirica*, the site should be selected in salt-affected areas with a salt content lower than 0.6–0.8 %, and the site preparation method of two-plowing-two-harrowing should be employed. Finally, a planting density of 2505/ha is best, and container seedlings planted in the rainy season, or the direct planting of bare-root seedlings for forestation, are good strategies.

References

Ashraf M (1994) Salt-tolerance of pigeon pea (*Cajanus cajan* (L.) *Millsp.*) at three growth stages. Ann Appl Biol 124(2):153–164

Chen SL, Li JK, Fritz E et al (2002) Sodium and chloride distribution in roots and transport in three poplar genotypes under increasing NaCl stress. For Ecol Manage 168(1–3):217–230

Gulzar S, Khan MA, Ungar IA (2001) Effect of temperature and salinity on the germination of *Urochondra setulosa*. Seed Sci Technol 29:21–29

Gupta GN (1987) Salt tolerance in some tree species at seedling stage. Indian For 112(2):101–112

Gupta GN, Mohan S, Prasad KG (1987) Salt-tolerance of selected tree seedlings. J Trop For 3(3):217–227

Jug A, Hofmann-Schielle C, Makeschin F (1999) Short-rotation plantations of balsam poplars, aspen and willows on former arable land in the Federal Republic of Germany II. Nutritional status and bioelement export by harvested shoot axes. For Ecol Manag 121(1–2):67–83

Kovda VA (1983) Loss of productive land due to salinazation. Ambio 7(2):91–93

Malcolm E, Sumner RN (1998) Sodic soils-distribution, properties, management, and environmental consequences. Oxford University Press, New York

Niknam SR, McComb J (2000) Salt tolerance screening of selected Australian woody species—a review. For Ecol Manage 139(1–3):1–19

Tabatabai MA (1994) Soil enzymes. In: Weaves RW, Angel GS, Bottomley PS (eds) Methods of soil analysis. Part 2, Microbiological and biochemical properties, vol 5, SSSA Book series. Soil Science society of America, Madison, pp 775–833

Zhang JF (2003) Advances of research on the mechanism of plant salinity tolerance and breeding of salt-tolerant plants. World For Res 16(2):16–22

Zhang JF (2004) Agroforestry and its application in amelioration of saline soils in eastern China coastal region. For Study China 6(2):27–33

Chapter 14
Tree Species Selection for Soil Salinity Resistance

Abstract In the Yellow River delta region, high soil salt content is one of the most important limited factors for the growth of trees, especially in riverbank areas, as on sloped land evaporation is to a certain extent a lot stronger, and soluble salts exist in the deep soil profile and easily travel to the soil surface. Hence, selecting suitable tree species is vital to implement ecological restoration in the region. In order to select suitable tree species, physiological parameters such as proline content, the ratio of bound water and free water, and relative conductivity in leaves were experimentally applied in the present reported study to probe the capability of trees to resist salinity stress. Based on the test results, the tree species *R. hispida*, *F. velutina*, *C. korshinskii*, *N. sibirica*, *T. chinensis*, *R. typhina*, *S. matsudana*, *A. fruticosa,* and *A. altissima* were found to possess salt tolerance, to various extents. Many studies have shown that proline content is linked with environmental stresses such as drought, cold, flooding, and salinity. Although to begin with most attention in the literature was paid to the role of proline as a compatible osmolyte and osmoprotectant, its other roles in stress tolerance have gradually received more interest. When plants grow in an adverse environment, proline content will usually increase. Similar results are reported in the present study.

Keywords Physiological indicator • Salinity tolerance • Tree species selection • Yellow River delta

The Yellow River delta (YRD) region is rich in natural resources, from which the crude oil, natural gas, chemical, crude salt, manufacturing, food processing, pulp, electricity, and bio industries all benefit (Zhang et al. 2004a). From the 1960s to 1990s, there was a large emphasis on oil exploitation and the utilization of other natural resources, and as a result, the natural environment received little protection. From Table 7.1, it can be seen that annual mean precipitation in the YRD region is 542.3 mm, but annual mean evaporation is 1962.1 mm. Under these conditions, drought often occurs, and soil salinization naturally follows (Zhang 2004;

J. Zhang, *Coastal Saline Soil Rehabilitation and Utilization Based on Forestry Approaches in China*, DOI 10.1007/978-3-642-39915-2_14,

Zhang 2005). In terms of future climate and soil conditions, the most difficult question in the context of ecological restoration efforts relates to the selection of tree species to tolerate drought and saline soil. In the present study, in order to examine this issue, some tree species were selected, and some indicators related to soil salinity stress were analyzed.

14.1 Materials and Methods

14.1.1 Site Description

The site used for the study is situated in Lijin County, Shandong Province, which is about 30 km from the mouth of the Yellow River. The soil salinity there is rather serious because of the dry weather, frequent winds, and low rainfall. In spring, white salt crusts usually appear on the soil surface, and because of the varied morphology, topography, and land use, the distribution of saline soil is highly variable. On average, soil salinity is about 0.87 %, with a maximum of 1.19 % and minimum of 0.16 %. The constituents of the soil salts are Cl^{-1} (59.48 mmol/kg), SO_4^{2-} (4.52 mmol/kg), $K^{+}+Na^{+}$ (42.85 mmol/kg), Mg^{2+} (13.63 mmol/kg), and Ca^{2+} (10.89 mmol/kg). Vegetation coverage is low, usually less than 35 %, with some halophyte plants growing sparsely, such as *Artemisia capillaris* Thunb., *Oenothera biennis*, *Phacelurus latifolius*, *Limonium sinensis*, *Ephedra sinica*, *Leymus chinensis*, *Artemisia capillaris*, *Apocynum venetum*, *Suaeda* spp., *Aeluropus* spp., *Phragmites australis*, and *Cynodon dactylon (L) pers* (Zhang et al. 2004b). Thus, soil and water erosion occurs badly when it rains, which affects the land quality and wildlife habitat of the region.

14.1.2 Plant Materials

In this study, one-year-old seedlings that had been propagated from seeds were sourced from Lijin State-owned Seedling Nursing Farm. For *Robinia hispida* L., the average seedling height was 2.1 m and the basal diameter 0.89 cm; while for *Fraxinus velutina* Torr., the mean seedling height was 1.87 m and the basal diameter 0.77 cm. Both initial stand spacings were 2×3 m.

14.1.3 Sampling and Analysis Methods

In order to examine the salinity-tolerance capabilities of the trees, a field survey was conducted in different specific stands with varied age classes and sites in the YRD region. When a tree grew well (survived and showed regular growth increments), whatever the level of soil salinity, it meant the tree could tolerate that salinity. When the tree grew badly in a specific field, especially when the tree tip began to wither, it indicated that the soil salt concentration was at the limit for successful tree growth. Accordingly, soil samples were collected from specific sites, and the salt content was measured. Meanwhile, leaves of specific trees were collected to analyze proline content, free-water content, and relative conductivity.

Proline was measured as described by Troll and Lindley (1995): 100 mg of frozen plant material was homogenized in 1.5 ml of 3 % sulphosalicylic acid, and the residue was removed by centrifugation. Then, 100 μl of the extract was reacted with 2-ml glacial acetic acid and 2-ml acid ninhydrin (1.25-g ninhydrin warmed in 30-ml glacial acetic acid and 20-ml 6 M phosphoric acid until dissolved) for 1 h at 100 °C, and the reaction was then terminated in an ice bath. The reaction mixture was extracted with 1 ml toluene. The chromophore-containing toluene was warmed to room temperature, and its optical density was measured at 520 nm. The amount of proline was determined from a standard curve in the range of 20–100 μg.

For measuring the relative conductivity (RC), the leaves were placed in a beaker containing 10 mL of distilled water for 1 h at room temperature, whereupon the conductivity of the water in the beaker was determined using a DSS 11 conductivity meter. The beaker was then placed in a boiling water bath for 10 min before being cooled to room temperature. The conductivity was then measured again, and the conductivity in the distilled water was measured as the control. The RC was calculated as the ratio of the difference between conductivities after 1 h versus the control to the difference between conductivity after boiling versus the control. The sampled leaves were also taken from the tree.

Relative water content of the leaves (RWC) was measured by the saturated weighing method (Lin and Ehleringer 1982). The samples were collected from the top of trees, and fresh leaves (fully expanded and sunlit) were weighed immediately and then incubated at 105 °C to deactivate the leaves, after which they were dried at 85 °C to constant weight. RWC was calculated according to the method of Dai et al. (2000). Free water content was measured using a reflective meter (2WAT). The bound water content and free water content are expressed as

$$\text{Bound water content } (\%) = \text{water content in fresh weight } (\%) - \text{free water content } (\%)$$

$$\text{Free water content } (\%) = B\,(B_2 - B_1) \,/\, \left(B_2 \times W_f\right) \times 100$$

where B is 60 % sugar liquid weight; B_1 is the sugar concentration at room temperature (20 °C); B_2 is the sugar concentration after the samples were immersed at 20 °C for 5 h; and W_f is fresh weight. Bound water was calculated using the above formula (Zhang et al. 1980).

Seedlings of the tree were planted in sites in April 2005. In November of that year, stand height and basal diameter for specific tree species were measured through a stand inventory; the standard area of a plot was 20 × 20 m. Three soil samples were collected from each plot, and the profile was divided into 0–20 cm, 20–40 cm, and 40–60 cm. Soil samples were air dried and sieved to < 2 mm. The pH was determined in distilled water at a soil/solution ratio of 1:2.5 with a potentiometric glass electrode (Zhang et al. 2008).

Soil salinity was measured using an electric conductivity meter DSS 11 (Singh and Thompson 1992). Soil organic matter was measured by the potassium dichromate method; total N by the semimicro Kjeldahl method; available P by the molybdenum blue colorimetry method through extraction with buffered 0.5 M NaHCO3; and available K by the flame photometry method through extraction with buffered 1 M NH4-acetate (Marcar et al. 1999; Singh et al. 1991).

All statistical analyses were performed using SAS and Excel software.

14.2 Results and Analysis

In the YRD region, high soil salt content is one of the most important limiting factors for tree growth. Hence, selecting suitable tree species is vital in order to implement ecological restoration projects in the region (Shields et al. 1995). Trees were chosen and their specific physiological features were analyzed. The results are shown in Table 14.1.

Many studies have shown that proline content is related to environmental stresses such as drought, cold, flooding, and salinity. Although initially most attention was focused on the role of proline as a compatible osmolyte (Bohnert et al. 1995) and osmoprotectant (Zhu 2000, 2002), its other roles in stress tolerance have gradually received more interest. When plants grow in an adverse environment, proline content will usually increase (Iyer and Caplan 1998; Liu and Zhu 1997). In this context, the effect of exogenous proline and/or stress on the expression of ubiquitin, antioxidative enzymes, and dehydrins was examined. If proline simulates the effect of salinity stress on these proteins in non-stressed plants and mitigates the negative effects in stressed plants, then it might be possible to confer its effect at least partially by inducing the stress-adaptive responses (Strizhov 1999; Yoshiba et al. 1997). That is why proline content is considered as one of the important indicators and is often measured when plants are under environmental stress (Okuma et al. 2000).

From the data presented in Table 14.1, it can be seen that for the 19 species grown in the YRD region, proline content varied greatly, being highest in *Nitraria sibirica* (340.49 μg/g) and lowest in *Toona sinensis* (20.84 μg/g). However, it can also be seen that for shrubs such as *Nitraria sibirica*, *Lycium chinense*, *Tamarix chinensis*, *Elaeagnus angustifolia*, and *Amorpha fruticosa*, the proline content was higher than in trees. This was reasonable and expected and was the same as when they were actually grown in the field. According to this single indicator, the order in terms of trees enduring salinity is *Rhus typhina*, *Fraxinus velutina* (see Fig. 14.1), *Robinia hispida*, *Melia azedarach* (see Fig. 14.2), *Ulmus pumila*, *Salix matsudana*, *Sophora*

Table 14.1 Physiological indicators of salinity tolerance for different tree species

Species	Salinity tolerance (%)	Proline content (μg/g/dw)	Free water content (%)	Bound water content (%)	BW/FW	Relative conductivity (%)
Nitraria sibirica	>1.5	340.49 ± 1.33	30.08 ± 0.65	28.92 ± 0.84	0.96	58.36 ± 1.37
Ulmus pumila	0.3–0.55	29.43 ± 0.66	34.7 ± 1.31	25.3 ± 0.57	0.73	44.92 ± 0.74
Ailanthus altissima	0.3–0.4	23.51 ± 0.86	37.2 ± 0.89	32.8 ± 1.24	0.88	47.57 ± 0.98
Tamarix chinensis	>1.5	68.41 ± 1.32	28.36 ± 0.85	26.64 ± 1.06	0.94	49.46 ± 0.59
Robinia pseudoacacia	0.3–0.55	25.88 ± 0.68	34.37 ± 1.35	28.63 ± 0.59	0.83	56.33 ± 1.44
Robinia hispida L.	0.3–0.5	32.57 ± 0.84	33.71 ± 0.78	27.29 ± 0.64	0.81	59.37 ± 0.67
Lycium chinense	0.4–0.7	110.54 ± 1.31	34.28 ± 0.76	28.72 ± 1.07	0.84	64.55 ± 0.82
Caragana korshinskii Kom.	0.3–0.6	35.74 ± 0.69	30.75 ± 0.93	28.25 ± 0.68	0.92	62.39 ± 0.59
Rhus typhina	0.25–0.4	56.13 ± 0.95	33.7 ± 0.81	30.3 ± 1.07	0.90	60.24 ± 1.46
Melia azedarach	0.3–0.5	30.22 ± 0.61	32.46 ± 0.85	27.54 ± 0.62	0.85	38.91 ± 0.72
Fraxinus velutina	0.3–0.55	33.17 ± 0.65	34.48 ± 0.54	31.52 ± 0.96	0.91	54.25 ± 0.61
Sophora japonica	0.3–0.4	27.25 ± 1.25	28.67 ± 1.32	25.33 ± 0.85	0.88	50.63 ± 1.42
Elaeagnus angustifolia	>1.0	57.63 ± 0.47	31.44 ± 0.61	28.56 ± 1.03	0.91	34.47 ± 0.58
Amorpha fruticosa	0.4–0.7	48.37 ± 0.76	34.72 ± 0.94	29.28 ± 0.66	0.84	52.74 ± 0.75
Salix matsudana	0.3–0.4	27.65 ± 0.83	34.67 ± 0.97	29.33 ± 1.25	0.84	48.55 ± 0.83
Toona sinensis	0.2–0.35	20.84 ± 0.79	35.63 ± 1.36	28.37 ± 0.86	0.79	32.66 ± 1.22
Morus alba	0.3–0.45	25.42 ± 0.81	35.13 ± 0.72	27.87 ± 0.97	0.79	40.72 ± 0.95
Ginkgo biloba	0.3–0.4	26.93 ± 1.46	35.37 ± 1.27	25.63 ± 1.34	0.72	45.84 ± 0.64
Campsis grandiflora	0.3–0.4	28.42 ± 0.68	33.88 ± 0.91	26.12 ± 1.06	0.77	51.45 ± 1.04

Fig. 14.1 Seedlings growing of *Fraxinus velutina* in Shandong (Photograph taken by Zhang Jianfeng)

Fig. 14.2 Seedlings growing of *Melia azedarach* in Shandong (Photograph taken by Zhang Jianfeng)

Fig. 14.3 Growing of *Lycium chinense* in Shandong (Photograph taken by Zhang Jianfeng)

japonica, *Ginkgo biloba*, *Robinia pseudoacacia*, *Ailanthus altissima*, *Morus alba*, and *Toona sinensis*. The result was accordant with soil salinity analysis. In view of this, it was concluded that proline content, as an important indicator of salinity tolerance, is useful and practical.

In plant tissue, there is both bound water and free water. The former is bound with substances such that it is difficult to freeze and evaporate, while the latter is concerned with metabolism; when free water in a tree is relatively high, its metabolism is active and the tree grows well. Bound water, free water, and their ratio have been found to be closely related to tree growth and salinity hardiness (Zhang 2008). The higher the ratio of bound water to free water, the stronger the tree's resistance to stress (Naidoo 1987). Among the 19 tree species tested, we found that the ratio was highest for *N. sibirica* (0.96), followed by *T. chinensis* (0.94), *C. korshinskii* (0.92), and *F. velutina* (0.91). For the other trees, the same rule more or less applied: if the ratio was higher, the resistance to salinity was stronger. It is clear that the ratio of bound water and free water can efficiently reflect the resistance of trees to salinity.

Relative conductivity is related to membrane permeability, which is crucial for plants in order to resist environmental stress (Kramer 1983). When trees are grown under salinity stress, the cell membrane first becomes injured, and then permeability increases. Some experiments have shown that RC could be linked with the capability of plants to resist external stress, that is, the higher the value of RC, the stronger the capability of the plant (Qureshi and Barrett 1998; Slavik 1974). As Table 14.1 indicates, some tree species, such as *L. chinense* (64.55) (see Fig.14.3),

C. korshinskii (62.39), *R. typhina* (60.2), *R. hispida* (59.37), and *N. sibirica* (58.3), had high values of RC, and they were capable of resisting salinity to varying degrees. However, there was an exception: *T. chinensis*. Its RC was 49.46, which was lower than some trees, such as *R. pseudoacacia* (56.33), *F. velutina* (54.25), and *S. japonica* (50.63). However, in fact, *T. chinensis* was stronger than those species in terms of salinity tolerance. This was perhaps due to different trees having been at different growth stages when the leaves were collected, meaning that at that time RC could not exactly reflect the tree's capability in resisting environmental stress.

14.3 Conclusions

In summary, the results of this study indicated that the more salt tolerant a tree is, the greater the values of the parameters analyzed (Gupta et al. 1986, 1987). Following the experiments, it is clear that selecting suitable tree species is a vital process. As part of this, understanding and analyzing certain physiological parameters—as was carried out in the present reported work—such as proline content, the ratio of bound water and free water, and relative conductivity, are equally important for examining the capabilities of trees to resist salt stress. Although not all the data were consistent with the capabilities of those trees studied here, generally speaking the three indicators were found to be effective, accessible, and practical. Based on experimental results, the species *R. hispida, F. velutina, C. korshinskii, N. sibirica, T. chinensis, R. typhina, S. matsudana, A. fruticosa,* and *A. altissima* were chosen and planted in the YRD area.

Additionally, selected species should also be able to withstand a degree of anticipated inundation, provide year-round protection, have the capacity to become well established under sometimes adverse soil conditions, and have root, stem, and branch systems capable of resisting erosive flows (Johnson and Stypula 1993; Shields et al. 2000). Furthermore, locally available native species should be used (Schiechtl and Stern 1994, 1996), and other important factors, including rapid initial growth, the ability to reproduce, and disease/pest resistance, should be considered.

References

Bohnert HJ, Nelson DE, Jensen RG (1995) Adaption to environmental stress. Plant Cell 7:1099–1111

Dai XB, Zhai HQ, Zhang HS et al (2000) Effect of drought stress on photosynthetic rate and carbonic anhydrase activity of rice leaf. Acta Phytophysiologica Sinica 26(2):133–135

Gupta GN, Prasad KG, Mohan S (1986) Salt-tolerance in some tree species at seedling stage. Indian For 112(2):101–113

Gupta GN, Mohan S, Rasad KGP (1987) Salt-tolerance of selected tree seedlings. J Trop For 3(3):217–227

Iyer S, Caplan A (1998) Products of proline catabolism can induce osmotically regulated genes in rice. Plant Physiol 116:203–211
Johnson AW, Stypula JM (1993) Guidelines for bank stabilization projects in the riverine environments of King County. King County Department of Public Works, Surface Water Management Division, Seattle
Kramer PJ (1983) Water relations of plants. Academic, New York, pp 325–326
Lin ZF, Ehleringer JR (1982) The effects of light, temperature, water vapor pressure deficit and carbon dioxide on photosynthesis in Papaya. Acta Phytophysiologica Sinica 8(4):363–372
Liu J, Zhu JK (1997) Proline accumulation and salt-stress-induced gene expression in a salt-hypersensitive mutant of *Arabidopsis*. Plant Physiol 114:591–596
Marcar N, Ismail S, Hossain A (1999) Trees, shrubs and grasses for saltlands, Australian Centre for International Agricultural Research, Canberra
Naidoo G (1987) Effects of salinity and nitrogen on growth and water relation in the mangrove *Avicennia marina* (Forsk.) Vierh. New Phytologist 107(2):317–325
Okuma E, Soeda K, Tada M et al (2000) Exogenous proline mitigates the inhibition of growth of *Nicotiana tabacum* cultured cells under saline conditions. Soil Sci Plant Nutr 46:257–263
Qureshi RH, Barrett-Lennard EG (1998) Saline agriculture for irrigated land in Pakistan: a handbook. Australian Centre for International Agricultural Research, Canberra
Schiechtl HM, Stern R (1994) Water bioengineering techniques for watercourse bank and shoreline protection. Osterreichischer Agrarverlag, Klosterneuburg
Schiechtl HM, Stern R (1996) Ground bioengineering techniques for slope protection and erosion control. Blackwell, London
Shields FD Jr, Bowie A, Cooper CM (1995) Control of streambank erosion due to bed degradation with vegetation and structure. Water Res Bull 31(3):475–489
Shields FD Jr, Knight SS, Cooper CM (2000) Warmwater stream bank protection and fish habitat: a comparative study. Environ Manag 26(3):317–328
Singh S, Thompson FB (1992) Salt-tolerance in some tropical tree species. Boiset Forets des Tropiques 234:61–67
Singh K, Yadav JSP, Singh V (1991) Tolerance of trees to soil salinity. J Indian Soc Soil Sci 39(3):549–556
Slavik B (1974) Methods of studying plant water relations. Academia (Czechoslovak Academy of Sciences), Prague
Strizhov N (1999) Different expression of two P5CS genes controlling proline accumulation during salt-stress requires ABA and is regulated by ABI1 and AXR2 in *Arabidopsis*. Plant J 12:557–569
Troll W, Lindley JA (1995) A photometric method for the determination of praline. J Biol Chem 215:655–660
Yoshiba Y, Kiyosue T, Nakashima K et al (1997) Regulation of levels of proline as an osmolyte in plants under water stress. Plant Cell Physiol 38:1095–1102
Zhang JF (2004) Agroforestry and its application in amelioration of saline soils in eastern China coastal region. For Study China 6(2):27–33
Zhang JF (2005) Causes of wetland degradation and ecological restoration in Yellow River delta region. For Study China 7(2):15–18
Zhang JF (2008) Principles and technologies of saline soil ecological amelioration. China Forestry Press, Beijing
Zhang XZ, Chen FY, Wang RF (1980) Plant physiological technology. Liaoning Technology Press, Shenyang, pp 5–60
Zhang JF, Xing SJ, Zhang XD (2004a) Principles and practice of forestation in saline soil in China. Chinese For Sci Technol 3(2):62–70
Zhang JF, Xing SJ, Zhou JX, et al (2004b) Study on cultural technologies and salt-resistance of *Nitraria sibirica* in coastal areas with serious salt-affected soil. In: Proceedings of the world engineers convention 2004, Vol F-A, Resources and Energy. Shanghai, China, pp 281–285

Zhang JF, Jiang JM, Xing SJ (2008) Planting techniques of *Tamarix chinensis* and its effect on saline soil remediation. In: Proceedings of the 2nd international conference on bioinformatics and biomedical engineering. Shanghai, China

Zhu JK (2000) Genetic analysis of plant salt tolerance using *Arabidopsis thaliana*. Plant Physiol 124:941–948

Zhu JK (2002) Salt and drought stress signal transduction in plants. Annu Rev Plant Biol 53:247–273

Chapter 15
Effect of Tree Planting on Preventing Riverbank Soil Erosion

Abstract At present, the reach downstream of the Yellow River is particularly badly eroded, having vertical banks up to 3–8 m high. The use of vegetation for bank protection is most effective and hence some trees were planted in the riverbank area. The results of a stand inventory implemented over a year showed that these trees grew well, especially *Lycium chinense*; the highest was 1.73 m (mean 1.45 m) and the largest basal diameter was 1.52 cm (mean 1.21 cm). After 3 years, the trees were planted in riverbanks, and soil samples were collected and analyzed to evaluate the effect of trees on soil amelioration. The results indicated that for wild land (ck), the mean organic matter (OM) was 6.75 g kg^{-1}, whereas for the stand of *Robinia hispida* and *Fraxinus velutina*, the mean OM was 7.42 and 7.32 g kg^{-1}, respectively. On average, OM was increased 9.03 % and 7.79 %, respectively. Moreover, the mean soil salinity was 1.99 g kg^{-1} for wild land, whereas for the stands of *R. hispida* and *F. velutina* it was 0.50 and 1.26 g kg^{-1}, respectively. In general, the salinity was decreased 298 % and 57.9 %, respectively. Hence, it was concluded that planting trees could prevent soils from salinization and protect riverbanks from degradation.

Keywords Riverbank • Soil amelioration • Soil erosion • Yellow River Delta

Approximately 2,000 ha of newly formed land are created each year in the Yellow River Delta (YRD) region by sediments carried by the river. This soil is mainly composed of fine sand and is characteristically young with a high groundwater table, low fertility, and a tendency toward secondary salinization and desertification (Zhang et al. 2008). In the YRD region, the average groundwater table is generally at a depth of 2–3 m, while along the coastline it is only at depths of 0.5–1.5 m. As a result of the encroachment of seawater and high evaporation, mineral concentration in the soil is usually 10–40 g L^{-1} but can reach 200 g L^{-1} in extreme cases (Zhang et al. 2002, 2010).

J. Zhang, *Coastal Saline Soil Rehabilitation and Utilization Based on Forestry Approaches in China*, DOI 10.1007/978-3-642-39915-2_15,

15.1 Introduction

The YRD is rich in natural resources. Crude oil, natural gas, chemical industry, and crude salt, as well as mechanical, food processing, pulp, electricity, and biological industries, are of primary importance (Zhang et al. 2004a). From the 1960s to 1990s, emphasis was placed on oil exploitation and the use of other natural resources. Thus, natural ecosystems were not regularly or effectively protected. Because the YRD is formed by silts from the Yellow River, the river course increases each year. Hence, the riverbank is gradually growing (Zhang 2005). In the YRD region, the river is severely eroded because of the erosion of sandy soils along the river course. The Yellow River in the delta region is currently higher in elevation than the surrounding areas (Zhang et al. 2004b). Downstream reaches are particularly eroded, with vertical banks reaching a height of 3–8 m. This makes it difficult to approach the water's edge, and in places where the river has been undercut and is susceptible to collapse, the bank represents a danger to people (Abramson et al. 2002). Therefore, enhancing riverbank protection, especially by building a riparian forest buffer zone, is vital for the stabilization of these areas (Gray and Leiser 1992; Gray and Sotir 1996).

Although the annual mean precipitation is 542.3 mm, the annual mean evaporation is 1,962.1 mm. Under these conditions, drought and soil salinization frequently occur (Zhang 2004). In consideration of possible future climate change and extensive soil erosion, one of the most difficult issues in riverbank restoration is selecting tree species that are tolerant of drought and high soil salinity. Based on previous experience and practice, several tree species were selected and planted along the YRD riverbank and plant growth and effects on soil properties were measured.

15.2 Materials and Methods

15.2.1 Site Description

This study area was located in Lijin County, Shandong Province, China, about 30 km from the mouth of the Yellow River.

In this area, soil salinity is high owing to xeric conditions, frequent winds, and low rainfall. In spring, a white salt crust usually appears on the slopes of the riverbank, and because of the sloping landform, the distribution of soil salts varies along the riverbank. Soil salinity generally averages 0.87 %, 1.19 %, and 0.16 % at the top, middle, and bottom of the slopes, respectively. Soil salts are mainly composed of 59.48 mmol kg^{-1} Cl^-, 4.52 mmol kg^{-1} SO_4^{2-}, 42.85 mmol kg^{-1} $K^+ + Na^+$, 13.63 mmol kg^{-1} Mg^{2+}, and 10.89 mmol kg^{-1} Ca^{2+}. Vegetation cover is low, usually less than 35 %. Some halophyte plants grow sparsely here, including *Artemisia capillaris* Thunb., *Oenothera biennis*, *Phacelurus latifolius*, *Limonium*

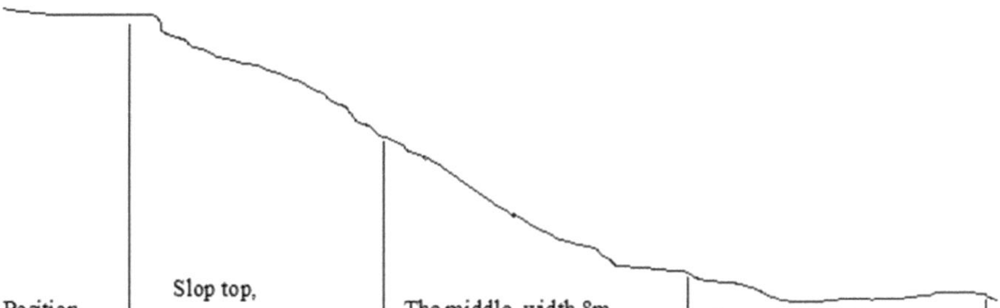

Position	Slop top, width 4m	The middle, width 8m	Bottom, width 6m
Veget-ation	*P. latifolius*, *L. chinensis*, vegetation coverage>60%	*Aeluropus* spp., salt crusts appear, vegetation coverage <35%	*Suaeda spp., P. australis*, vegetation coverage >60%
Soils	0-40cm, Soil salinity 0.87%, soil moisture 13.8%, pH8.2	0-40cm Soil salinity 1.19%, soil moisture 17.0 %, pH7.9	0-40cm Soil salinity 0.16%, the highest 0.33%, soil moisture 19.2 %, pH8.3

Fig. 15.1 Site description

sinensis, *Ephedra sinica*, *Leymus chinensis*, *Artemisia capillaries*, *Apocynum venetum*, *Suaeda* spp., *Aeluropus* spp., *Phragmites australis*, and *Cynodon deatylin* (L.) pers (Zhang et al. 2004b; Zhang and Jiang 2010). Given the sparse vegetation, soil and water erosion are severe during rainfall events, which affect land quality and wildlife habitats along the river channel. Figure 15.1 shows a cross-section of the sloping riverbank illustrating typical vegetation and soil properties.

15.2.2 Plant Materials

In this study, all seedlings used for planting the riverbank were 1 year old, propagated by seeds, and obtained from the Lijin State-owned Seedling Nursing Farm. Four species were chosen for this study. The shrubs *Lycium chinense* and *Tamarix chinensis* were planted as plantlets and the stems were cut. Basal diameters for the two species averaged 0.68 and 0.82 cm, respectively. The initial height of *Robinia hispida* L. seedlings averaged 2.1 m, and the basal diameter averaged 0.89 cm. The mean height of *Fraxinus velutina* Torr. seedlings was 1.87 m, and the basal diameter was 0.77 cm. Initial stand spacing was 2 m × 3 m.

15.2.3 Sampling and Analysis

To assess the salinity tolerance of the tree species, field surveys were conducted in specific stands with varied age classes and at different sites within the YRD region. Trees that had high survival and growth rates were assumed to be tolerant of the salinity of the soil in which they grew. Trees with low growth rates in a specific field, or those with an apex that had begun to wither, indicated that the soil salinity was at a maximum for tree growth. Soil samples were collected in specific sites, and salt content was measured.

Trees were planted in the riverbanks in April 2007. In November of that year, stand height and basal diameter for specific species were measured. The area of the plots was 20 m × 20 m. In October 2009, soil properties were analyzed in stands of *Robinia hispida* L. and *Fraxinus velutina* Torr. Adjacent, unplanted areas served as controls.

Three soil samples were collected in each plot, and the profile was divided into 0–20, 20–40, and 40–60 cm sections. Soil samples were air dried and sieved to <2 mm. pH was determined in distilled water at a soil/solution ratio of 1:2.5 with a potentiometric glass electrode (Zhang et al. 2008).

Soil salinity was measured with an Electric Conductivity Meter DSS 11 (Singh and Thompson 1992). Soil organic matter (OM) was measured by the potassium dichromate method, and total nitrogen was determined with the semi-micro Kjeldahl method. Available phosphorus was assessed by the molybdenum blue colorimetry method by extraction with buffered 0.5 M $NaHCO_3$, and available potassium was analyzed by the flame photometry method by extraction with buffered 1 M NH_4^- acetate (Marcar et al. 1999; Singh et al. 1991).

All statistical analyses were performed with SAS and Excel software.

15.3 Results

15.3.1 Tree Growth

Based on results of the field survey and analysis of physiological indicators, certain tree species were selected and planted in the riverbank area, including *R. hispida*, *F. velutina*, *T. chinensis*, and *L. chinense*. Shoots of seedlings were cut, and only the basal section of the main stem was retained when planting. Inventories of stands were performed in December 2009, and results are presented in Figs. 15.2 and 15.3.

As the figures indicate, after one growing season height increment and diameter increment of all species were relatively high. *L. chinense* responded especially well with a highest height increment of 1.73 m (mean 1.45 m), and the largest basal diameter was 1.52 cm (mean 1.21 cm). High growth indicates that the four selected species were well adapted to riverbank soil conditions despite the adverse environment (see Fig. 15.4).

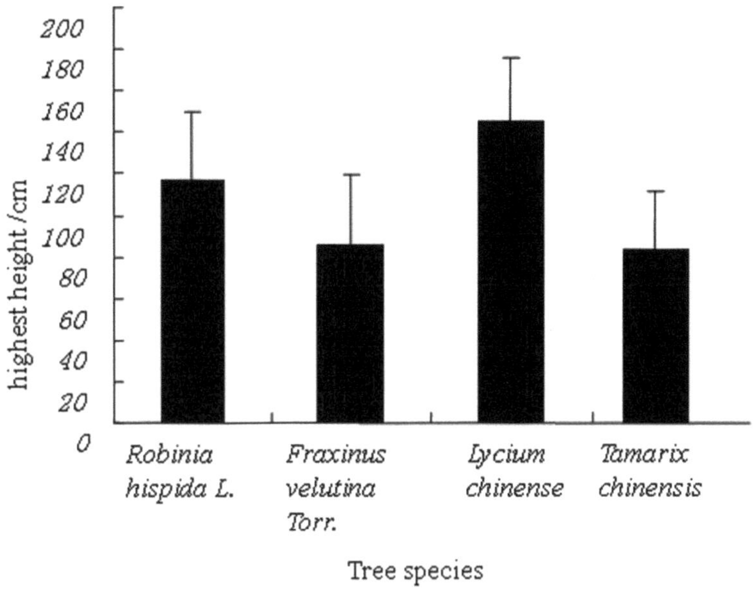

Fig. 15.2 Height increment of different trees

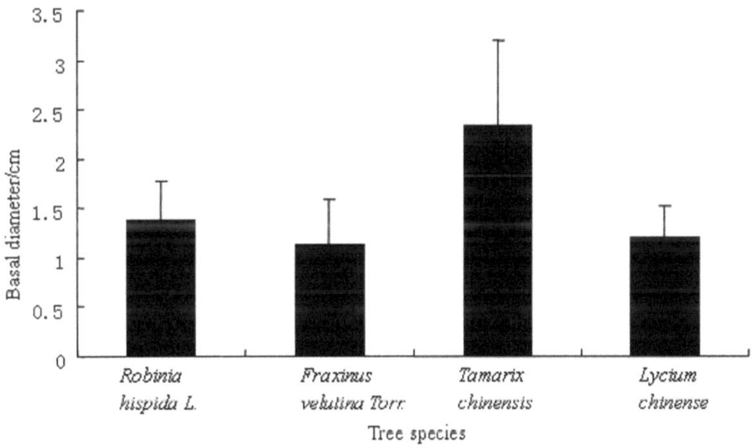

Fig. 15.3 Diameter increment of the four species

15.3.2 Effects of Planted Seedlings on Soil Properties

To investigate the ability of planted trees to ameliorate soil conditions, soil samples were collected from forested sloping land 3 years after the trees were planted. Trees improved soil nutrition, and soil salinity decreased compared with the wild land (Table 15.1).

Fig. 15.4 Growing of *Robinia hispida* and *Tamarix chinensis* at the riverbank (Photograph taken by Zhang Jianfeng)

Table 15.1 Soil salinity and change in nutrition

Stands	Profile (cm)	Organic matter ($g\ kg^{-1}$)	Total nitrogen ($g\ kg^{-1}$)	Effective nitrogen ($mg\ kg^{-1}$)	Effective phosphorus ($mg\ kg^{-1}$)	Effective potassium ($mg\ kg^{-1}$)	Soil salinity ($g\ kg^{-1}$)
Robinia hispida	0–20	9.47	0.784	43.52	11.03	157.95	0.34
	20–40	6.35	0.636	41.74	10.14	131.04	0.42
	40–60	6.43	0.371	33.49	9.01	124.71	0.75
	Mean	7.42	0.597	39.58	10.06	137.90	0.50
Fraxinus velutina	0–20	12.06	0.467	42.71	11.17	81.01	0.98
	20–40	5.59	0.433	31.23	9.79	55.22	1.12
	40–60	4.31	0.269	27.32	8.95	47.83	1.69
	Mean	7.32	0.389	33.75	9.97	61.35	1.26
Wild land (ck)	0–20	10.65	0.321	35.69	10.48	74.38	1.58
	20–40	4.83	0.232	17.39	9.61	41.25	2.12
	40–60	4.77	0.161	14.77	9.18	47.04	2.28
	Mean	6.75	0.238	22.62	9.76	54.22	1.99

Soil OM is a basic parameter that indicates changes in soil nutrition. The mean OM for wild land (ck) was 6.75 $g\ kg^{-1}$, while the mean OM for stands of *R. hispida* and *F. velutina* were 7.42 and 7.32 $g\ kg^{-1}$, respectively. Results were similar for the other indicators. The average OM of *R. hispida* and *F. velutina* stands increased 9.03 % and 7.79 %, respectively, compared with wild land. In addition, the increase in vegetation cover that was associated with reforested riverbanks reduced soil evaporation.

15.4 Discussion

In China, the Yellow River is called the "Mother River," which implies its crucial role and significance in the region. As a result of climate change and increasing human disturbance, land degradation in banks is becoming worse, which may lead to deterioration of soil conditions, a reduction in vegetation coverage, and even dam damage. Hence, in view of the present state of the riverbanks of the Yellow River in the delta region, use of vegetation for bank protection and environmental improvement is most effective when combined with rational structural components.

The correct selection of the vegetation species is vital to establish biological shelter belts in riverbanks. These species should be able to withstand the level of anticipated inundation, providing year-round protection, have the ability to become well established under sometimes adverse soil conditions, and have root, stem, and branch systems capable of resisting erosion (Johnson and Stypula 1993; Shields et al. 2000). Other important plant attributes include rapid initial growth, ability to reproduce, and resistance to disease and insects. Locally available native species should be used whenever possible (Schiechtl and Stern 1994, 1996).

In addition to planting trees, existing natural vegetation should be protected on undisturbed portions of riverbanks, especially in areas subject to flooding. Overhanging trees on riverbanks that are undercut need to be removed, but roots and stumps should be left intact. These trees should be replaced with new trees that will provide additional structural support. Alternatively, the undercut zone around the tree base can be filled and stabilized with riprap (Polster 1998).

Furthermore, previous research has shown that root systems are able to hold soil, and that litter from planted species is cycled back into soils to enrich OM (Panchaban et al. 1989; Pinsley and Swift 1986).

In the current study, by improving the physical and chemical conditions of the soil, planted seedlings consequently reduced soil salt concentration. Mean soil salinity was 1.99 g kg^{-1} for wild land and 0.50 g kg^{-1} and 1.26 g kg^{-1} for stands of *R. hispida* and *F. velutina*, respectively, which represent decreases of 298 % and 57.9 %. Hence, planting trees can prevent soil salinization and reduce riverbank degradation (Qureshi and Barrett-Lennard 1998; Zhang et al. 2011).

Moreover, based on these results, *R. hispida* may be better than *F. velutina* in protecting riverbanks in saline areas. In addition to fast growth, this is a leguminous species that fixes nitrogen. Therefore, it may be especially effective for preventing soil degradation (see Fig. 15.5).

15.5 Conclusions

In this study, various tree species were selected based on the results of field surveys at the study site and were planted along the riverbank. Planted tree species exhibited relatively good growth. *L. chinense*, in particular, responded well with the largest

Fig. 15.5 Growing of *Robinia hispida* (*left*) and *Fraxinus velutina* (*right*) (Photograph taken by Zhang Jianfeng)

height increment of 1.73 m (mean 1.45 m), and the biggest basal diameter of 1.52 cm (mean 1.21 cm). Three years after the trees were planted soil samples were collected and analyzed to evaluate the effects of trees on soil amelioration. Mean OM in adjacent wild land (control) was 6.75 g kg^{-1}, while OM in stands of *R. hispida* and *F. velutina* was 7.42 g kg^{-1} and 7.32 g kg^{-1}, respectively, or increases of 9.03 % and 7.79 %. Moreover, mean soil salinity was 1.99 g kg^{-1} in wild land and 0.50 g kg^{-1} and 1.26 g kg^{-1} in stands of *R. hispida* and *F. velutina*, respectively. This represents salinity decreases of 29.8 % and 57.9 %.

In places where severe bank erosion occurs, stands of full-grown trees provide little protection for stabilizing riverbanks apart from roots binding the soil. Shrubs can potentially provide much better protection. With their extensive root systems, shrubs are better able to hold soil in place and reduce water velocities. In addition to structural support, shrubs provide important functions, such as improving soil conditions, protecting riverbanks from erosion, and minimizing sediment and associated nutrient impacts downstream. They also protect tree trunks from damage caused by ice breaks and help prevent the formation of strong eddies around large trees during flood events. These results imply that planting trees can prevent soil salinization and protect riverbanks from degradation in the YRD region.

References

Abramson LW, Thomas SL, Sharma S et al (2002) Slope stability and stabilization measures, 2nd edn. Wiley, New York

Gray DH, Leiser A (1992) Biotechnical slope protection and erosion control. Van Nostrand Reinhold, New York

Gray DH, Sotir RB (1996) Biotechnical and soil bioengineering slope stabilization – a practical guide for erosion control. Wiley, New York, pp 286–289

Johnson AW, Stypula JM (1993) Guidelines for bank stabilization projects in the riverine environments of King County. King County Department of Public Works, Surface Water Management Division, Seattle

Marcar N, Ismail S, Hossain A (1999) Trees, shrubs and grasses for salt lands: an annotated bibliography. Australian Centre for International Agricultural Research, Canberra
Panchaban S, Katawatin R, Srisataporn P (1989) Effect of salinity on growth of fast growing trees. Thai J Agric Sci 2:91–99
Pinsley RT, Swift MJ (1986) Amelioration of soil by trees – a review of current concepts and practices. Marlborough House, London
Polster DF (1998) Introduction to soil bioengineering, participant manual-forest renewal BC. Polster Environmental Services, Duncan
Qureshi RH, Barrett-Lennard EG (1998) Saline agriculture for irrigated land in Pakistan: a handbook. Australian Centre for International Agricultural Research, Canberra
Schiechtl HM, Stern R (1994) Water bioengineering techniques for watercourse bank and shoreline protection. Osterreichischer Agrarverlag, Klosterneuburg
Schiechtl HM, Stern R (1996) Ground bioengineering techniques for slope protection and erosion control. Blackwell, London
Shields FD Jr, Knight SS, Cooper CM (2000) Warmwater stream bank protection and fish habitat: a comparative study. Environ Manage 3:317–328
Singh S, Thompson FB (1992) Salt-tolerance in some tropical tree species. Bois et Forêsts des Tropiques 234:61–67
Singh K, Yadav JSP, Singh V (1991) Tolerance of trees to soil salinity. J Indian Soc Soil Sci 3:549–556
Zhang JF (2004) Agroforestry and its application in amelioration of saline soils in eastern China coastal region. Forest Study China 2:27–33
Zhang JF (2005) Causes of wetland degradation and ecological restoration in Yellow River Delta region. Forest Study China 2:15–18
Zhang JF, Jiang JM (2010) Soil salinization and ecological remediation by planting trees in China. In: Proceedings of 2010 international conference on mechanic automation and control engineering. Wuhan, China, pp 1349–1352
Zhang JF, Song YM, Xing SJ (2002) Saline soil amelioration and forestation techniques. J Northeast Forest Univ 6:124–129
Zhang JF, Xing SJ, Zhang XD (2004a) Principles and practice of forestation in saline soil in China. China Forest Sci Technol 2:62–70
Zhang JF, Xing SJ, Zhou JX et al (2004b) Study on cultural technologies and salt-resistance of *Nitraria sibirica* in coastal areas with serious salt-affected soil. In: Proceedings of the World engineers' convention 2004. Resources and energy, vol F-A. Shanghai, China, pp 281–285
Zhang JF, Jiang JM, Xing SJ (2008) Planting techniques of *Tamarix chinensis* and its effect on saline soil remediation. In: Proceedings of the 2nd international conference on bioinformatics and biomedical engineering, IEEE, New York, pp 4259–4262
Zhang JF, Chen GC, Xing SJ et al (2010) Carbon sequestration of black locust forests in the Yellow River Delta region, China. Int J Sustain Dev World Ecol 17:475–480
Zhang JF, Chen GC, Xing SJ (2011) Water shortages and countermeasures for sustainable utilisation in the context of climate change in the Yellow River Delta region. Int J Sustain Dev World Ecol 18:177–185

Chapter 16
Function of Black Locust Plantation on Enhancing Soil Properties

Abstract The Yellow River currently debouches north of the Shangdong peninsula. The primary cause of this rapid shift has been the heavy load of silt carried by the Yellow River. As the current slackens near the sea, much of this is redeposited, building up the riverbed and sooner or later forcing the water to run elsewhere. Should a flood occur, the river may break out of the levees into the surrounding lower flood plain and adopt a new course. This has occurred historically about once every 100 years. In modern times, considerable effort has been made to strengthen levees and control floods. The results have shown that soil salinity was reduced 48.64 %; nutrients such as organic matter, total nitrogen and effective nitrogen, total phosphorus and effective phosphorus, and effective potassium were increased 37.32 %, 19.72 %, 53.89 %, 29.71 %, 19.85 %, and 13.10 %, respectively. Soil enzyme activity was higher for forested land, which implied that growing trees could ameliorate soil conditions in the riverbank and help prevent soil erosion to some extent. Therefore, it is concluded that the developing black locust forest ecosystem in the Yellow River Delta region is significant in terms of this adverse and fragile environment. For this ecosystem, the forest is the framework for and main body preventing soil erosion and enhancing riverbank stabilization. Moreover, with stand growing, its function of C sequestration will gradually increase, which implies the importance of tree plantation in the district not only for riverbank protection.

Keywords Black locust • Plantation • Soil improvement

The Yellow River rises on the northern slopes of the Bayankela Mountains of Qinghai and traverses nine provinces or autonomous regions including Qinghai, Gansu, Sichuan, Ningxia, Inner Mongolia, Shaanxi, Shanxi, Henan, and Shandong in China. The river falls 4,450 m over a length of 5,465 km, draining an area of 795,000 km^2 (Zhang and Li 2002).

J. Zhang, *Coastal Saline Soil Rehabilitation and Utilization Based on Forestry Approaches in China*, DOI 10.1007/978-3-642-39915-2_16,

A notable feature of the river is the largest loess deposit in the world. The Yellow River far exceeds any of the world's large rivers in terms of annual sediment transport. Because of the large quantity of silts, the delta extends into the sea at a rate of 2.2 km per year with a land accretion of 3,240 ha (Zhang 2008; Zhang et al. 2011). Thus, it has the highest silt content in the world, with an average annual sediment load of 1.049 billion t and a mean silt concentration of 25.5 kg m^{-3} (Zhang 2004).

16.1 Introduction

In its lower reach, over a length of almost 800 km, the river becomes a broad meandering channel. As a result of its heavy load of silt the Yellow River becomes a depositing stream, and consequently at the estuary the Yellow River Delta (YRD) is formed by silt. As it deposits part of its carried burden of soil in its bed in stretches where it flows slowly, these deposits elevate the riverbed that flows between natural levees in its lower reaches, and the bed rises above the surrounding land at a rate of 1 m every 10 years. At present, the build-up of silt in the lower reaches of the river is severe, with 400 million t filling the riverbed every year (Zhang 2005; Zhang and Jiang 2010). The silt has built up a new riverbed, known by locals as the "second hanging river," on top of the original riverbed, on the lower reaches from Kaifeng, Henan province, to the river mouth.

The river at present debouches north of the Shangdong peninsula. The primary cause of this rapid shift has been the heavy load of silt carried by the Yellow River. As the current slackens near the sea, much of this is redeposited, building up the riverbed and sooner or later forcing the water to run elsewhere. Should a flood occur, the river may break out of the levees into the surrounding lower flood plain and adopt a new course. This has historically occurred about once every 100 years. In modern times, considerable effort has been made to strengthen levees and control floods (Sun and Zhang 2006; Zhang 1997). Moreover, the YRD itself is subject to the effects of natural disasters, such as storm surges in the winter season and steady coastal erosion in the delta areas now devoid of sedimentation, and excessive siltation in the active parts of the delta.

The economic development of the delta is now calling for a stabilization of the delta branches by "river greening." A first and foremost development requirement is to harnass the river course. It should be possible to keep the river between its embankments, the sedimentation problem needs to be taken care of, and the riverbank must be protected from soil erosion (Schiechtl and Stern 1996; Zhang et al. 2008, 2010).

Therefore, planting trees along the riverbank in the YRD is vital and significant. In terms of the local adverse environment, black locust (*Robinia pseudoacacia*) was chosen as the main tree species for afforestation in the region. It is a type of leguminous plant that grows fast and is capable of withstanding a certain degree of soil salinity (Pinsley and Swift 1986; Zhang et al. 2002). Its flowers are edible and its leaves are fine fodder. To probe its multi-functions, we researched the important effects of the tree on soil rehabilitation.

16.2 Methods

16.2.1 Study Area and Site Descriptions

The test plot is situated at 118°33′–119°20′E, 37°35′–38°12′N, in the northeast of Dongying, Shandong Province, the People's Republic of China, which faces the Bohai Sea in the North and borders Laizhou Bay in the East (Zhang and Li 2002). It has a warm-temperate continental monsoon climate, with a dry and windy spring, a hot and rainy summer, a cool and clear autumn, and a cold, dry, and snowless winter. This area has an average annual temperature of 11.9 °C, a frost-free period of about 210 days, an average annual precipitation of 592 mm, a mean annual evaporation of 1,962 mm, and an average annual relative humidity of 68 %. Water of the Yellow River has a pH of 8.0–8.3, a total hardness of 2.16–5.56, and a mineral intensity of 0.2–0.6 g L^{-1}. The average ground-water table is generally 2–3 m, while along the coastline it is only 0.5–1.5 m. As a result of the encroachment of seawater and high evaporation, mineral concentration in the soil is usually 10–40 g L^{-1} and in extreme cases is 200 g L^{-1} (Zhang et al. 2004b).

The soil type in the area is mainly fluvo-aquic soil accompanied with saline soil, cinnamon soil, and brown earth, and along the area of estuary is mainly offshore saline soil. The soil in the inner-bank area is generally saline soil and in the area outside the bank is mainly clay and loam (Wu and Zhang 1995). The fluvo-aquic soil is mainly distributed on the river sediments and influenced by ground water activities; the fertility of the soil varies according to the source and quality of the sediments; the soil shows generally alkalescency, a high content of organic matter (OM), and is suitable for farming.

The main flood season is from July to October, when there is heavy rainfall in the middle and lower reaches. From October to mid-December, the river flow is steady. Then, from the end of December to March it will be ice bound, during which the water level often rises drastically owing to poor drainage from ice blocking the river channel (Zhang et al. 2004a, b).

The measured plot lies in the Hekou district in Dongying. The stand area is 10 ha of pure block locust that was established in 1985. Understory vegetation was mainly *Pelargonium hortorum* Bailey., *Rubia cordifolia* Linn., *Digitaria sanguinalis* (Linn.) Scop., *Erysimum cheiranthoides* Linn., *Salsola collina* Pall., *Hippochaete ramosissimun* Desf., *Viola philippica* Cav., *Cynosurus viridis* (Linn.) Beauv., *Imperata cylindrical* (Linn.) Beauv., *Meconopsis kanashiroi* Kitam., *Afzelia nigrum* Linn., *Meconopsis hedinii* Ostenf. et Pauls., and *Argyranthemum coronarium* Linn. (Sun and Zhang 2006).

Fig. 16.1 Field inventory at the stand of *Robinia pseudoacacia* (Photograph taken by Zhang Jianfeng)

16.2.2 Experimental Details

The test plantation was established in the spring of 1985, some plantlets were recruited when others died or were destroyed in the second year. Seedlings were 1 year old from the Lijin State-owned Seedling Nursery, the mean height was 1.6 m and the basal diameter was 0.5–0.8 cm. The top shoot of the plantlet was cut to 30–50 cm high when planted. The initial spacing of the stand was 2.5 m × 3 m.

In October of 2006, the stand was inventoried by random sampling, and the inventory plot area was 20 m × 20 m (see Fig. 16.1). The area surrounding the wild land (without the trees) in the riverbank was the control. For every plot, soil profiles were created and soil samples were collected according to 0–20, 20–40, 40–60, 60–80, and 80–100 cm. All plant material and roots were removed and the samples were stored at 4 °C for up to 5 days before analyses took place. Soil samples were air dried and sieved to <2 mm.

The pH was determined in distilled water at a soil:solution ratio of 1:2.5 with a potentiometric glass electrode (Marcar et al. 1999). Analysis of Na^+ was conducted after digestion by HNO_3 using an inductively coupled argon plasma emission spectrophotometer (Jobin-Yvon JY 48) (Naidoo 1987).

Soil salinity was measured by an Electric Conductivity Meter DSS 11(Panchaban et al. 1989).

Soil OM of the soil sub-sample was determined by potassium dichromate oxidation revised as per the $K_2Cr_2O_7$-voluming outer-heating method. Results were expressed as mass fraction '%' (carbon mass divided by dry soil mass) (Yang et al.

1998). While total nitrogen was measured by the semi-micro Kjeldahl method; available phosphorus was determined by the molybdenum blue colorimetry method by extraction with buffered 0.5 M $NaHCO_3$; and available potassium was analyzed by the flame photometry method by extraction with buffered 1 M NH^{4-} acetate (Singh et al. 1991; Xu et al. 2004).

Soil enzyme activity was assayed by Tabatabai (1994), Yu (2001), and Zhou (1987).

All analyses were carried out in triplicate.

All statistical data were performed with SAS and Excel software.

16.3 Results

16.3.1 Soil Salinity

In the YRD region, higher soil salinity was the most adverse limiting factor for trees to grow. After black locust plantations were established, plant coverage in the bank was increased and accordingly soil evaporation was reduced. Consequently, salts existing in deep soil went to the topsoil slowly, which lead to soil salinity in forested land being lower than that in the wild land where no trees were planted (see Table 16.1). The soil salinity was indicated by electric conductivity (EC).

Table 16.1 shows that soil salt concentration in forested land was decreased for every soil profile. Especially for 0–60 cm, the effect of trees on soil salinity alleviation was obvious and significant, i.e., EC was 2.34–5.83, while for the control the value was 6.64–12.21; at deeper levels, the effect was weaker. This implied that the black locust forest could adapt to the natural environment in the bank and improve soil conditions. In fact, field soil salinity was quite serious, even in the spring white salt crusts appeared on the soil surface, and accordingly few plants could grow well (Gong 1986; Zhang 2008). Thus, the black locust plantation was crucial for protecting the riverbank, because once the stand was established, other plant communities would grow and improve the entire environment (Abramson et al. 2002; Gray and Leiser 1992; Gray and Sotir 1996).

Table 16.1 Electric conductivity under different land-use systems (ds cm^{-1})

Soil profile (cm)	Forested land	Control
0–20	2.34 ± 0.74	6.64 ± 0.86
20–40	2.55 ± 0.82	10.27 ± 0.71
40–60	5.83 ± 1.03	12.21 ± 0.91
60–80	8.27 ± 0.46	14.84 ± 0.69
80–100	12.23 ± 0.93	16.82 ± 0.58
Mean	6.24 ± 1.85	12.15 ± 2.49

Table 16.2 Na^+ percentage under different land-use systems and pH

Soil profile (cm)	Forested land (%)	Control (%)
0–20	0.142 ± 0.841	0.154 ± 0.683
20–40	0.159 ± 0.644	0.217 ± 0.744
40–60	0.181 ± 0.736	0.238 ± 0.548
60–80	0.192 ± 0.837	0.259 ± 0.642
80–100	0.304 ± 0.561	0.298 ± 0.716
pH	7.41 ± 0.847	8.05 ± 0.688

16.3.2 Soil Na^+ Content and pH

It is well known that for coastal salt-affected soils, the main components of salinity are Na^+, K^+, Ca^{2+}, Mg^{2+}, and Cl^- (Gupta et al. 1987; Zhang 2004). Generally speaking, Na^+ content to some extent reflects the physiological and chemical features of the soil (Qureshi and Barrett-Lennard 1998). Hence, Na^+ was analyzed and the results, including the pH values, are given in Table 16.2.

Table 16.2 shows that Na^+ percentage was lower in the forest soils than in wild soils. It has been shown that Na^+ content is related to soil structure and nutrition absorption, such as porosity, bulk density, and osmotic potential, as well as plant growing (Bohnert et al. 1995; Wu and Zhang 1995). This implied that the black locust plantation could prevent soil salinization, and the results were consistent with those in Table 16.1. Moreover, the pH value of the two land-use systems also exhibited the function of trees. However, it was worth mentioning that in an 80–100 cm profile, Na^+ percentage was higher in the forested soil than in the wild soil. This might suggest that the primary Na^+ percentage was more or less the same between the different land-use types, but because of tree growing, much more Na^+ congregated around the roots together with water and the other nutrients (Zhang et al. 2008). This question still needs to be examined further.

On the one hand, soil pH was concerned with soil properties, e.g., texture, soil solution acidity, or basicity, and, on the other hand, soil pH was related to plant absorption, i.e., when the pH was higher, some ions were uptaken more; when the pH fell, the other ions were easily absorbed (Naidoo 1987; Polster 1998). Nevertheless, for good plant growing, a certain pH was required and this was generally about 5.5–6.2 or slightly acidic. This creates the greatest average level for availability for all essential plant nutrients. Extreme fluctuations of a higher or lower pH can cause deficiency or toxicity of nutrients (Marcar et al. 1999). Normally, macronutrients tend to be less available in soils with a low pH and micronutrients tend to be less available in soils with a high pH.

In this case, the pH in forested land was 7.41, which might be adjusted by trees through leaf litter rotting or root secretion, and the value was nearly suited to tree growing.

Table 16.3 Nutrients contents under different land-use systems

Nutrients	Forested land	Control	Change (%)
Total nitrogen (%)	0.071 ± 0.63	0.057 ± 0.76	19.72
Effective nitrogen (mg kg^{-1})	80.12 ± 1.05	36.94 ± 0.86	53.89
Total phosphorus (%)	0.138 ± 0.73	0.097 ± 0.81	29.71
Effective phosphorus (mg kg^{-1})	20.15 ± 0.66	16.15 ± 0.75	19.85
Effective potassium (mg kg^{-1})	179.28 ± 0.91	155.79 ± 1.72	13.10
Organic matter (%)	1.14 ± 1.42	0.71 ± 0.83	37.72

16.3.3 Soil Nutrition

In the YRD, downstream riverbank soil nutrition was rather poor because of soil erosion and salinization (Sun and Zhang 2006; Wu and Zhang 1995). Thus, it was one of the vital parameters for examining land-use potential. Here, some soil nutrients were measured from a profile of 0–40 cm; the results are shown in Table 16.3.

From Table 16.3 it shows that the soil nutrition state with a plantation growing was better than that without trees, especially the parameter of OM, which usually could be considered as the general indicator of nutrition. In this case, OM reached 1.14 % owing to leaf litter from the plantation-enriching soils. Moreover, in terms of effective nitrogen, the function of black locust trees on soil improvement was also ideal, i.e., the content in the soil with trees was increased 37.72 % in land without a forest. Nitrogen is a major component of proteins, hormones, chlorophyll, vitamins, and enzymes essential for plant life and transfer of energy (Xu et al. 2004). Nitrogen metabolism is a major factor in stem and leaf growth (vegetative growth), helping plants increase seed and fruit production, and improving the quality of leaf and forage crops. Deficiencies can reduce yields, causing yellowing of the leaves and stunting growth (Li et al. 2003). Obviously, it was understood that fertilizer application was impossible in the region, thus nitrogen often came from the air. While black locust trees as a type of legume could get nitrogen from the atmosphere so that the soil fertilization improved, trees would consume some for growth.

Like nitrogen, phosphorus is an essential part of the process of photosynthesis. It is involved in the formation of all oils, sugars, and starches, helps with the transformation of solar energy into chemical energy, helps with proper plant maturation, withstands stress, ensures rapid growth, and encourages blooming and root growth (Cookson 2002; Li 1994). Although the plantation could not provide phosphorus to soils, by reinforcing soil conditions and soil microbiological properties, tree growing was helpful for superphosphate to transform.

Potassium is necessary for the formation of sugars, starches, carbohydrates, protein synthesis, and cell division in roots and other parts of the plant. It helps to adjust water balance and improves stem rigidity and resistance to cold temperatures

Table 16.4 Soil enzyme activity under different land-use systems

Soil enzyme activity	Forested soil	Control
Invertase (0.1 mol/L$Na_2S_2O_3$ mL)	0.072	0.051
Phosphatase (P_2O_5 mg)	79.69	32.63
Catalase (0.1 mol/L1/5$KMnO_4$ mL)	0.136	0.098

(Bohnert et al. 1995). In the YRD region, potassium in soils was not deficient because of the fluvo-aquic soil (Wu and Zhang 1995). Hence, Table 16.3 shows that the value of potassium content in both lands was nearly equal (179.28 and 155.79 mg kg^{-1}).

16.3.4 Soil Enzyme Activity

It is clear that enzymes such as invertase, phosphatases, and catalases play a key role in soil nutrient transformation. When trees are planted, soil enzyme activity should vary (Browman and Tabatabai 1978). The results are shown in Table 16.4.

In Table 16.4, we found that although the main functions varied among the three enzymes, the results were similar, i.e., enzyme activity in forested land was higher than in wild land. This means that with tree growing soil conditions were ameliorated to some extent, which not only embodied soil fertility, but also soil microbial activity (Li 1994).

Invertase is involved in carbon cycling; its value is concerned with carbon/nitrogen and soil nutrition. Plant residues, which usually make the largest contribution of organic material to soils, contain both labile and more recalcitrant compounds. The labile compounds are usually soluble and easily degraded, such as simple sugars like glucose and amino acids (Eivazi and Tabatabai 1977). The more recalcitrant components are typically structural biopolymers such as starch, cellulose, and lignin. Because extracellular soil enzymes are directly responsible for the initial processing of detrital carbon and organic-bound nutrients, it is expected that enzyme activity would be higher in land with trees than land without trees.

Phosphorus is one of the least available of all essential nutrients in the soil and its concentration is generally below that of many other macronutrients (Grierson et al. 2000). Many soils around the world are deficient in inorganic phosphorus, and even in fertile soils available inorganic phosphorus rarely exceeds 10 μmol L^{-1} (Li 1994), thus phosphorus fertilizers are necessary to ensure plant productivity. Extracellular phosphatase enzymes catalyze the hydrolysis of organic phosphorus compounds into forms of inorganic phosphorus available for uptake by both plants and microbes (Hojeong and Chris 1999) and contribute to the extracellular degradation of complex polymeric molecules to more readily assimilable monomers. Phosphatase enzymes are inducible under conditions of low inorganic phosphorus (Zhou 1987) and this activity varies considerably among plant species and, for mycorrhizal roots, among the species of fungal symbionts (Cookson 2002). In Tables 16.3 and 16.4, we can see the result was consistent for enzyme activity and inorganic phosphorus level in different soils.

Catalase is a highly conserved heme containing an antioxidant enzyme known for its ability to degrade hydrogen peroxide into water and oxygen (Tabatabai 1994). Our results describe a novel function for catalase in the metabolism of endogenous substrates and in the action of carcinogens and chemopreventative agents.

16.4 Discussion

As previously mentioned, the silt built up a new riverbed, known as the "second hanging river" on top of the original riverbed, on the lower reaches of the Yellow River. Thus, to keep the river between its embankments the sedimentation problem needs to be resolved. The first and most important assignment is to protect the riverbank from soil erosion (Shields et al. 1995, 2000). In this regard, the choice of suitable tree species and establishing plantations in the riverbank is vital (Schiechtl and Stern 1994).

By a soil profile analysis and a stand inventory, the results confirmed that the black locust tree was adapted to the local climate and soil conditions. Through planting trees the soil salinity was reduced by 48.64 %; nutrients such as OM, total nitrogen and effective nitrogen, total phosphorus and effective phosphorus, and effective potassium were increased 37.32 %, 19.72 %, 53.89 %, 29.71 %, 19.85 %, and 13.10 %, respectively, and soil enzyme activity was higher for forested land.

Some research has shown that soil enzyme activity could be considered as the indicator of soil nutrition (Eivazi and Tabatabai 1977). Hence, in this test, three enzymes were assayed. A previous assumption has been that catalase functions only to remove excess cellular hydrogen peroxide and/or to metabolize small-molecular-weight electron donors such as ethanol (Tabatabai 1994). Further studies are needed to more precisely characterize the site and the reaction mechanism for the oxidase reaction in catalase, and to define the role of this enzymatic activity in cellular metabolism. Production of extracellular phosphatase enzymes is also dependent on environmental conditions, the physiological state of the plant, and the root type, root age, and position on the root (Hojeong and Chris 1999). The commonly observed inverse relationship between inorganic phosphorus in the soil solution and phosphatase activity suggests that measurement of specific enzyme activities can be used as indicators of relative nutrient limitation (Grierson et al. 2000). However, because the majority of enzyme activity may be abiotic, effects in the bulk soil and specifically at the root surface or within the microbial biomass may not be detectable.

16.5 Conclusions

The results show that soil salinity was reduced 48.64 %; nutrients such as OM, total nitrogen and effective nitrogen, total phosphorus and effective phosphorus, and effective potassium were increased 37.32 %, 19.72 %, 53.89 %, 29.71 %, 19.85 %, and 13.10 %, respectively, and soil enzyme activity was higher for forested land, which implies that tree growing could ameliorate soil conditions in the riverbank and help prevent soil erosion to some extent.

Therefore, it is concluded that the developing black locust forest ecosystem in the YRD region is significant in terms of the adverse and fragile environment. For this ecosystem, the forest is the framework and main body for preventing soil erosion and enhancing riverbank stabilization (Johnson and Stypula 1993; Shields et al. 2000). Moreover, with stand growing, its function of carbon sequestration will gradually increase, which implies the importance of plantation in the district not only for riverbank protection.

References

Abramson LW, Thomas SL, Sharma S et al (2002) Slope stability and stabilization measures, 2nd edn. Wiley, New York, pp 22–37

Bohnert HJ, Nelson DE, Jensen RG (1995) Adaption to environmental stress. Plant Cell 7:1099–1111

Browman MG, Tabatabai MA (1978) Phosphodiesterase activity of soils. Soil Sci Soc Am J 42:284–290

Cookson P (2002) Variation in phosphatase activity in soil: a case study. Agric Sci 7(1):65–72

Eivazi F, Tabatabai MA (1977) Phosphatases in soils. Soil Biol Biochem 9:167–172

Gong HZ (1986) Silviculture in saline soil. China Forestry Press, Beijing, pp 62–66

Gray DH, Leiser A (1992) Biotechnical slope protection and erosion control. Van Nostrand Reinhold, New York, pp 28–36

Gray DH, Sotir RB (1996) Biotechnical and soil bioengineering slope stabilization: a practical guide for erosion control. Wiley, New York, pp 286–289

Grierson PF, Comerford NB et al (2000) Non-destructive measurement of acid phosphatase activity in the rhizosphere using nitrocellulose membranes and image analysis. Plant Soil 218:49–57

Gupta GN, Mohan S, Prasad KG (1987) Salt-tolerance of selected tree seedlings. J Trop Forest 3(3):217–227

Hojeong K, Chris F (1999) Phosphatase and arylsulphatase activities in wetland soils: annual variation and controlling factors. Soil Biol Biochem 31:449–454

Johnson AW, Stypula JM (1993) Guidelines for bank stabilizaton projects in the riverine environments of King County. King County Department of Public Works, Surface Water Management Division, Seattle, pp 47–92

Li BW (1994) The correlation of phosphatase activity and organic phosphorus fractions and the effect of organic materials on it in Chao Soil. J Hebei Agric Univ 17(4):54–57

Li DP, Wu ZJ, Chen LJ et al (2003) Dynamics of urease activity in a long-term fertilized black soil and its affecting factors. Chin J Appl Ecol 14(12):2208–2212

Marcar N, Ismail S, Hossain A (1999) Trees, shrubs and grasses for Saltlands. Australian Centre for International Agricultural Research, Canberra, pp 62–128

Naidoo G (1987) Effects of salinity and nitrogen on growth and water relation in the mangrove *Avicennia marina* (Forsk.) Vierh. New Phytol 107(2):317–325
Panchaban S, Katawatin R, Srisataporn P (1989) Effect of salinity on growth of fast growing trees. Agric J 17(2):91–99
Pinsley RT, Swift MJ (1986) Amelioration of soil by trees: a review of current concepts and practices. Marlborough House, London, pp 22–47
Polster DF (1998) Introduction to soil bioengineering. Participant Manual-Forest Renewal BC. Polster Environmental Services, Duncan, pp 42–84
Qureshi RH, Barrett-Lennard EG (1998) Saline agriculture for irrigated land in Pakistan: a handbook. Australian Centre for International Agricultural Research, Canberra, pp 26–57
Schiechtl HM, Stern R (1994) Water bioengineering techniques for watercourse bank and shoreline protection. Osterreichischer Agrarverlag, Klosterneuburg, pp 46–78
Schiechtl HM, Stern R (1996) Ground bioengineering techniques for slope protection and erosion control. Blackwell Science, London, pp 15–48
Shields FD Jr, Bowie AJ, Coope CM (1995) Control of streambank erosion due to bed degradation with vegetation and structure. Water Resour Bull 31(3):475–489
Shields FD Jr, Knight SS, Cooper CM (2000) Warmwater stream bank protection and fish habitat: a comparative study. Environ Manage 26(3):317–328
Singh K, Yadav JSP, Singh V (1991) Tolerance of trees to soil salinity. J Indian Soc Soil Sci 39(3):549–556
Sun QX, Zhang JF (2006) Discussion on forestry sustainable development in the Yellow River Delta region. Chin Forest Sci Technol 5(4):63–67
Tabatabai MA (1994) Soil enzymes. In: Weaver RW, Augle JS, Bottomley PS (eds) Methods of soil analysis: 2. Microbiological and biochemical properties, vol 5, SSSA book series. SSSA, Madison, pp 775–883
Wu LH, Zhang SJ (1995) Rhizospheric effects of organic material on coastal saline soil in Huanghe River Delta I: effect on soil nutrient and enzyme activity. Chin J Appl Ecol 6(2):160–165
Xu FL, Liang YL, Zhang CE et al (2004) Effect of fertilization on cucumber growth and soil biological characteristics in sunlight greenhouse. Chin J Appl Ecol 15(7):1227–1230
Yang JQ, Wang ZZ, Guo CL (1998) Effect of organic matter on soil biological activity in warping field. Acta Agric Boreali-Sin 13(2):82–86 (in Chinese)
Yu QY (2001) Study on soil phosphatase activity and their influenced factors. J Anhui Agric Coll 15(4):5–8
Zhang JF (1997) Review of saline soil utilization. J Shandong Forest Sci Technol 4:22–25
Zhang JF (2004) Agroforestry and its application in amelioration of saline soils in eastern China coastal region. Forest Study China 6(2):27–33
Zhang JF (2005) Causes of wetland degradation and ecological restoration in Yellow River Delta region. Forest Study China 7(2):15 18
Zhang JF (2008) The principles and techniques on ecological rehabilitation of saline soils. China Forestry Press, Beijing, pp 24–46
Zhang JF, Jiang JM (2010) Soil salinization and ecological remediation by planting trees in China. In: Proceedings of 2010 international conference on mechanic automation and control engineering. Wuhan, China, pp 1349–1352
Zhang JF, Li XF (2002) Strategies to reclaim and ameliorate saline soil in the Yellow River Delta region. In: Chi Chang, Brian Dobing (eds) International conference on environmentally sustainable agriculture for dry areas for the 3rd millennium proceedings. Dobing Enterprises, Canada, pp 264–269
Zhang JF, Song YM, Xing SJ (2002) Saline soil amelioration and forestation techniques. J Northeast Forest Univ 6:124–129
Zhang JF, Xing SJ, Zhang XD (2004a) Principles and practice of forestation in saline soil in China. Chin Forest Sci Technol 3(2):62–70
Zhang JF, Xing SJ, Zhou JX, et al (2004b) Study on cultural technologies and salt-resistance of *Nitraria sibirica* in coastal areas with serious salt-affected soil. In: Proceedings of the world engineers' convention 2004, Resources and energy, vol F-A. Shanghai, China, pp 281–285

Zhang JF, Jiang JM, Xing SJ (2008) Planting techniques of *Tamarix chinensis* and its effect on saline soil remediation. In: Proceedings of the 2nd international conference on bioinformatics and biomedical engineering. IEEE, New York, pp 4259–4262
Zhang JF, Chen GC, Xing SJ et al (2010) Carbon sequestration of black locust forests in the Yellow River Delta region, China. Int J Sustain Dev World Ecol 17(6):475–480
Zhang JF, Chen GC, Xing SJ (2011) Water shortages and countermeasures for sustainable utilisation in the context of climate change in the Yellow River Delta region, China. Int J Sustain Dev World Ecol 18(2):177–185
Zhou LK (1987) Soil enzymology. Science Press, Beijing, pp 12–35

Chapter 17
Planting Black Locust (*Robinia pseudoacacia*) Forest as a Biomass Energy Resource

Abstract At present, the whole world is facing the challenge of energy shortage. Developing biomass energy resources is vital. However, in China, land area is limited and thus exploiting wild lands to plant fast-growing trees as a biomass energy resource is significant. There is an area of land (1,200,000 ha) in the Yellow River Delta region with rich natural resources; however, because of the adverse and fragile environment, few trees could grow there. Black locust (*Robinia pseudoacacia*) is the main tree species suited for afforestation. To analyze its growing state and biomass production, a black locust plantation was established and a field survey was conducted. On the basis of the results of a standard plot inventory and a stem disc analysis, the mean biomass of the black locust plantation was obtained: 166.85 t ha^{-1}. The result suggests that for biomass energy resource production, the trunks of the trees played an important role, and moreover, the function was enhanced with stand growing.

Keywords Black locust forest • Biomass energy resources • Yellow River Delta

It is well known that today the world is facing the challenge of energy shortage. Especially in China, with population growth and a booming economy, energy consumption is rapidly increasing. Therefore, the development of renewable energy as an alternative is crucial. In fact, humans have used biomass energy or bioenergy from organic matter for thousands of years, ever since people started burning wood to cook food or to keep warm. In fact, today, wood is still the largest biomass energy resource (Fung et al. 2002).

17.1 Introduction

China's use of biological energy to generate power has developed rapidly and the total installed capacity has reached 2.2 million kW, of which 1.7 million kW was from biogases co-generation and 500,000 kW was powered by agriculture

J. Zhang, *Coastal Saline Soil Rehabilitation and Utilization Based on Forestry Approaches in China*, DOI 10.1007/978-3-642-39915-2_17,

and forestry wastes, methane, direct burning of rubbish, and landfill gas. In 2007, China's fuel ethanol output reached 1.6 million t, of which 80 % used corn as the raw material. Based on the proportion 1:3.3, more than 4 million t of corn was consumed. It is estimated that China's annual fuel ethanol output will reach 5 million t in 2010 and the rate of use ethanol gasoline will be more than 50 % (Zhang 2008). Due to the outbreak of the financial crisis since 2008, the oil price falls from the top. The enthusiasm for fuel ethanol production also goes down. Hence the real production of fuel ethanol in 2011 is 1.77 million tons, while in 2012 it is 1.66 million tons in China. However, corn-based ethanol has stopped expanding production capacity because government policies have restricted the development of grain-based ethanol. Therefore, developing non-grain ethanol has become an inevitable trend.

There are many types of biomass, including wood, plants, residue from agriculture or forestry, and the organic component of municipal and industrial wastes. These can all be used as renewable energy sources (Stokes 1992). Today, many bioenergy resources are replenished through the cultivation of energy crops, such as fast-growing trees and grasses. These energy crops are called biomass feedstocks. The use of biomass feedstocks can also help increase profits for the agricultural industry. Biomass can be burned directly to produce steam for electricity production or manufacturing processes. Using biomass does not add to global warming as plants can use and store carbon dioxide (CO_2) when they grow. The CO_2 is released when the plant material is burned and other plants in turn consume the released CO_2. Using biomass closes this cycle of storing CO_2 (USDA Forest Service 2003), thus developing bioenergy that is carbon free and environment friendly.

However, where to plant the trees? In China, ensuring grain security is the most important issue. Consequently, farmlands are strictly protected and not permitted to change to another land use. Because of this situation, the only option is to exploit wild lands, such as saline soil or desert lands, by planting fast-growing trees as a biomass energy resource.

The Yellow River rises on the northern slopes of the Bayankela Mountains of Qinghai and traverses nine provinces or autonomous regions including Qinghai, Gansu, Sichuan, Ningxia, Inner Mongolia, Shaanxi, Shanxi, Henan, and Shandong in China. The river falls 4,450 m over a length of 5,465 km, draining an area of 795,000 km^2 (Zhang and Li 2002).

A notable feature of the river is the largest loess deposit in the world. The Yellow River far exceeds any of the world's large rivers in terms of annual sediment transport. Owing to the large quantity of silts, the delta extends into the sea at a rate of 2.2 km/year with a land accretion of 3,240 ha (Zhang 2008). However, these lands are hindered by climatic, soil, and environmental factors, e.g., drought, waterlogging, salinity, wind, and pollution. Thus, soil quality and productivity in the region is not compatible with economic development; furthermore, soil degradation will affect the ecological balance as well as environmental stability.

Apart from producing biomass feedstock, the establishment of forests in the region is also an effective method to overcome environmental problems, in view of the natural merits of woody plants (Zhang 2002). Based on in situ investigation and site classification, black locust (*Robinia pseudoacacia*) was chosen as the main

Fig. 17.1 Establishment of *Robinia pseudoacacia* stand in the Yellow River Delta (Photograph taken by Zhang Jianfeng)

tree species for afforestation (see Fig. 17.1). At present there is 6,000 ha black locust forest growing as a pure stand or a mixed stand in line with specific land-use systems (Song 2001). It is hoped that these trees could be a biomass feedstock.

Hence, the study on the characteristics of the trees in the Yellow River Delta (YRD) region is significant in view of the specific site conditions and the unique location, although no further tree species are grown there.

17.2 Description of Natural Conditions in the Region

The YRD is one of the biggest deltas in China, with an area of 1,200,000 ha. The scale of latitude and longitude is 37°21′–38°12′ N and 118°07′–119°10′ E, respectively. The delta is located in a temperate zone characterized by a monsoon climate.

The annual increment of newly formed land is 6,068 ha owing to sands depositing from 1974 to 1984 and 2,300 ha from 1984 to 2003. The area of salt-affected soil is 443,000 ha. Such soils are mainly composed of fine sands from the Yellow River with characteristics of young age, a high ground-water table, low fertility, and a tendency for secondary salinization and desertification. The average ground-water table is generally 2–3 m, while along the coastline it is only 0.5–1.5 m. Because of the encroachment of seawater and high evaporation, mineral concentration in the soil is usually 10–40 g/L and in extreme cases is 200 g/L (Zhang 2004).

There are altogether 393 plant species and variations in the region, and most of them are salt tolerant. The natural vegetation is broadleaf deciduous forest (mainly Hankow willow and Weeping willow), shrubbery (mainly Chinese tamarisk), and shore coppice. Besides the indigenous species, hundreds of new salt-tolerant species have been introduced in recent years and many of them grow well (Zhang 2002; Gong 1986).

The measured plot lies at the 11th Farm of Yellow River Delta Production Base in Dongying. The stand area is 100 ha of pure black locust that was established in 1985. Understory vegetation was mainly *Pelargonium hortorum* Bailey., *Rubia cordifolia* Linn., *Digitaria sanguinalis* (Linn.) Scop., *Erysimum cheiranthoides* Linn., *Salsola collina* Pall., *Hippochaete ramosissimun* Desf., *Viola philippica* Cav., *Cynosurus viridis* (Linn.) Beauv., *Imperata cylindrical* (Linn.) Beauv., *Meconopsis kanashiroi* Kitam., *Afzelia nigrum* Linn., *Meconopsis hedinii* Ostenf. et Pauls., and *Argyranthemum coronarium* Linn.

17.3 Materials and Methods

The test plantation was established in the spring of 1985, some plantlets were recruited when others died or were destroyed in the second year. Seedlings were 1 year old from the Lijin State-owned Seedling Nursery, the mean height was 1.6 m and the basal diameter was 0.5–0.8 cm. The top shoot of the plantlet was cut to 30–50 cm high when planted. The initial spacing of the stand was 2.5 m × 3 m.

In October of 2006, the stand was inventoried by random sampling, and the inventory plot area was 20 m × 20 m.

The stand biomass was measured by “standard stock law.” Based on the data of each stock recording, three standard stocks were determined, cut, and the biomass was gained separately. Moreover, discs of each trunk were sawed according to the normal rules (Vleeshouwers and Verhagen 2002).

All analyses were carried out in triplicate.

All statistical data were performed with SAS and Excel software.

Measurement was of the current stand annual biomass increment.

As mentioned above, discs of the trunk were collected during the standard stock inventory, and then the current annual increment could be achieved through stem analysis based on the following formula:

$$\begin{aligned} v &= v_1 + v_2 + v_3 + \cdots + v_n + v_{top} \\ &= g_1 \times l_1 + g_2 \times l_2 + g_3 \times l_3 + \cdots + g_n \times l_n + \frac{1}{3} g' \times l' \\ &= g_1 \times l_1 + l \sum_{i=2}^{n} gi + \frac{1}{3} g' \times l' \end{aligned} \tag{17.1}$$

Where

g_i – trunk disc area of the section No. i;
l – length of the section;
g' – basal area of the topwood;
l' – length of the topwood;
n – number of the section.

Accordingly, the current stand biomass increment could be computed based on the stand volume (Vleeshouwers and Verhagen 2002).

17.4 Results and Analysis

17.4.1 Standard Stock Biomass

According to diameter class, the standard stocks were selected and trunk discs were collected as well as the standard plot inventory. Consequently, the biomass of stocks was computed and the results are shown in Table 17.1.

Table 17.1 Biomass distribution in black locust standard stock

Stock no.	Part of tree	Biomass (kg)	Distribution (%)
1	Leaf	4.80	2.43
	Branch	43.35	22.11
	Stem	103.52	54.23
	Fine root	0.51	0.24
	Medium root	3.87	1.91
	Coarse root	37.09	19.08
Total		193.14	
2	Leaf	3.24	1.84
	Branch	37.39	21.45
	Stem	87.49	51.55
	Fine root	0.45	0.24
	Medium root	1.91	1.06
	Coarse root	41.23	23.86
Total		171.71	
3	Leaf	2.64	1.90
	Branch	33.59	24.41
	Stem	69.43	51.80
	Fine root	0.37	0.25
	Medium root	3.41	2.39
	Coarse root	26.27	19.25
Total		135.71	
Mean		166.85	

Table 17.2 Current annual biomass increment and net annual increment of the tree

Year	Current annual biomass increment (%)	Net annual biomass increment (%)	Current annual biomass increment (t ha^{-1})	Net annual biomass increment (t ha^{-1})
1	0.18	0.18	0.39	0.39
2	0.70	0.53	1.57	1.17
3	1.87	1.16	4.16	2.59
4	2.89	1.02	6.43	2.27
5	5.66	2.77	12.58	6.16
6	10.07	4.41	22.40	9.81
7	13.96	3.89	31.04	8.64
8	18.21	4.25	40.49	9.46
9	22.54	4.33	50.12	9.63
10	27.50	4.96	61.16	11.03
11	33.76	6.26	75.09	13.93
12	38.48	4.72	85.59	10.51
13	43.00	4.51	95.63	10.04
14	46.44	3.44	103.28	7.65
15	52.88	6.44	117.60	14.32
16	58.15	5.28	129.34	11.74
17	65.13	6.98	144.86	15.52
18	72.71	7.58	161.72	16.86
19	81.24	8.53	180.70	18.98
20	90.50	9.26	201.29	20.59
21	100.00	9.50	222.41	21.12
Total	100.00	100.00	222.41	222.41
Mean				10.59

Table 17.1 shows that the black locust could produce a great amount of biomass, and that each part of the tree produced different amounts, e.g., the stem was the main contributer, accounting for more than 50 % of the whole, the root also played an important role, especially coarse roots. Although different parts of the tree were used, generally speaking, the more biomass it the tree had the more bioenergy that was produced.

17.4.2 Current Annual Biomass Increment and Net Annual Biomass Increment of the Tree

To approach the dynamics of black locust growth, current annual biomass increment and net annual biomass increment of the tree were taken into consideration. The results are shown in Table 17.2.

It was concluded from Table 17.2 that the biomass of the stand increased with growth of the trees, although the increment varied in different ages. At the initial

Fig. 17.2 Growth of *Robinia pseudoacacia* in the Yellow River Delta (Photograph taken by Zhang Jianfeng)

stage (1–5 years), the annual biomass increment was around 1.4 t ha^{-1}. When the stand was in the fast-growing stage (6–14 years), the annual biomass increment was approximately 5–6 t ha^{-1}. When it reached a stable stage (above 15 years), the annual biomass increment was over 8 t ha^{-1}. This means that when the stand reached the age of 15 years, the biomass growth is at its peak, afterwards, the growing fells. Thus, the optimal time to harvest and gain the highest yields, is at this time. It suggests that the forest ecosystem of black locust has a great potential for biomass production (see Fig. 17.2).

The world is struggling to deal with climate change and energy shortage. CO_2 is one of the principal greenhouse gases and its concentration in the atmosphere has been steadily increasing (Bruce et al. 1999). The concentration of CO_2 in the atmosphere over the past 60 years has increased from 280 to 365 ppm (Eve et al. 2001). This increase in CO_2 concentration is expected to have serious environmental consequences and global warming is one such major consequence (Izaurralde et al. 2000).

Terrestrial ecosystems represent the second largest carbon reservoir and the carbon balance can be directly impacted by human activities such as agricultural management practices and land-use changes (Allmaras et al. 2000).

The use of biomass energy has the potential to greatly reduce our greenhouse gas emissions. Biomass generates about the same amount of CO_2 as fossil fuels, but every time a new plant grows, CO_2 is actually removed from the atmosphere. The net emission of CO_2 will be zero as long as plants continue to be replenished for biomass energy purposes.

Biomass (organic matter) can be used to provide heat, make fuel, and generate electricity. Wood, the largest source of bioenergy, has been used to provide heat for thousands of years. There are many types of biomass, including wood, plants, residue from agriculture or forestry, and the organic component of municipal and industrial wastes. These can all be used as renewable energy sources. In view of the adverse environment in the YRD region, planting black locust trees could produce a biomass of 166.85 t ha^{-1}, which is better than other tree species (Zhang 2008). Hence, the research on the biomass increment of the black locust forest ecosystem in the YRD region is significant as it is an adverse and fragile environment. Moreover, with stand growing, the function of biomass production will enhance gradually, which implies the importance of forestation in the district.

References

Allmaras RR, Schomberg HH, Douglas CL et al (2000) Soil organic carbon sequestration potential of adopting conservation tillage in U.S. croplands. J Soil Water Conserv 55(3):365–373

Bruce JP, Frome ME, Haites H et al (1999) Carbon sequestration in soils. J Soil Water Conserv 54:382–389

Eve MD, Paustian K, Follett R et al (2001) A national inventory of changes in soil carbon from National Resources Inventory data. In: Lal R, Kimble JM, Follett RF, Stewart BA (eds) Assessment methods for soil carbon. CRC Publishers, Boca Raton

Fung PYH, Kirschbaum MUF, Raison RJ et al (2002) The potential for bioenergy production from Australian forests, its contribution to national greenhouse targets and recent developments in conversion processes. Biomass Bioenergy 22:223–236

Gong HZ (1986) Silviculture in saline soil. China Forestry Press, Beijing, pp 62–66

Izaurralde RC, Rosenberg NJ, Lal R (2000) Mitigation of climate change by soil carbon sequestration: issues of science, monitoring, and degraded lands. Adv Agron 70:1–75

Song YM (2001) Study on salt-tolerance and major planting techniques of *Tamarix chinensis*. In: Liu MY (eds) The proceedings of international symposium on halophyte utilization and regional sustainable development of agriculture. Meteorology Press, Beijing, pp 68–72

Stokes BJ (1992) Harvesting small trees and forest residues. Biomass Bioenergy 2(1):131–147

USDA Forest Service (2003) A strategic assessment of forest biomass and fuel reduction treatments in Western States. General Technical Report RMRS-GTR-149. U.S. Department of Agriculture, Forest Service, Rocky Mountain Research Station, Fort Collins, 17p

Vleeshouwers LM, Verhagen A (2002) Carbon emission and sequestration by agricultural land use: a model study for Europe. Global Change Biol 8:519–530

Zhang JF (2002) Research on saline soil amelioration and forestation techniques. J Northeast Forest Univ 30(6):124–129

Zhang JF (2004) Agroforestry and its application in amelioration of saline soils in eastern China coastal region. Forest Study China 6(2):27–33

Zhang JF (2008) The principles and techniques on ecological rehabilitation of saline soils. China Forestry Press, Beijing, pp 24–46

Zhang JF, Li XF (2002) Strategies to reclaim and ameliorate saline soil in the Yellow River Delta region. In: Chi Chang, Brian Dobing (eds) International conference on environmentally sustainable agriculture for dry areas for the 3rd millennium proceedings, Dobing Enterprises, Canada, pp 264–269

Chapter 18
Responses of Plant Community Change on Wetland Degradation in Yellow River Delta Region

Abstract As a result of climate change, oil field development, and increased human activity, the wetlands of the Yellow River Delta region of China exhibit regional degradation. To investigate the mechanism of wetland degradation and biodiversity changes, we performed a vegetation survey and a soil property analysis, including soil enzyme activity, in the Yellow River Delta region. The results reveal that at the Imperata (*Imperata cylindrical*) community stage, the soil is slightly alkalized, and the soil surface salinity is generally less than 0.3 %. Under such conditions, Imperata can grow and develop well, with good soil structure, low soil bulk density, high porosity, high fertility, and high soil organic matter content, which are conducive to plant growth. If the vegetation is not destroyed, the soil remains at this stage or further improves. When the salinity increases, the soil gradually becomes unsuitable for Imperata growth, and halophytes begin to appear, forming a halophyte community stage. At this stage, soil salinity increases, surface salinity increases to 0.8 % or higher, and the soil has a more compact structure, higher soil bulk density, and lower porosity. When vegetation degrades to a *Suaeda glauca* community stage, soil salinity further increases and evolves into a severe saline-alkali soil, with significantly augmented soil salinity, enhanced salinization, compact soil texture, poor permeability, and lower soil fertility. Further degradation of wetlands leads to vegetation disappearance, leaving bare land with only sporadic *S. glauca*. The soil environment is extremely harsh, with a high salt content, the soil is compact and sterile, and the organic matter content is merely 2.4 g/kg. In such an environment, plants have difficulty growing. Thus, vegetation evolution and degradation of wetlands occur almost simultaneously. Soil salinity accelerates the reverse succession of phytocoenosium, which in turn shifts the physical and chemical properties of soil toward a bad direction for plant growth. The two factors complement each other and are closely related. Analysis of different stages of community succession showed that, along with the degradation succession of vegetation, the physical and chemical properties of soil cause the corresponding changes, with a gradually increased degree of soil salinity.

J. Zhang, *Coastal Saline Soil Rehabilitation and Utilization Based on Forestry Approaches in China*, DOI 10.1007/978-3-642-39915-2_18,

Keywords Biodiversity • Phytocoenosium • Wetland degradation • Yellow River Delta

The Yellow River Delta (YRD) region is formed by the continuous swing of the Yellow River tail. The parent material is Yellow River alluvial deposits, and the bottom is composed of marine sediment. It is one of three major river deltas in China, located at east longitude 118° 07′–119° 10′ and north latitude 37° 20′–38° 10′, with a total area of approximately 8,100 km^2. It is centered at Ninghai of the Kenli District, starting north from the Taoer River mouth to south at the Zimai River mouth. It forms a fan-shaped topography to the east, with an altitude of less than 15 m. The modern YRD region is the alluvial fan created by Yellow River alluvial effects since 1855 (Zhang et al. 2011). It is a marine sediment plain formed by large amounts of sand carried by the Yellow River to fill the depressed land of Bohai. The area has a warm temperate and continental monsoon climate, with four distinct seasons, sufficient sunshine, equal periods of rain and heat, and an average annual rainfall of 551.6 mm. The region is wide and open, with few people, and diverse landscapes such as flood heights, slopes, and a large-scale depression (Xu 1996). The site condition is complex, and the wetlands and swamps are widely distributed.

18.1 Introduction

The YRD region is one of the fastest land-building river deltas in the world, and approximately 20–23 km^2 of new silt land forms each year. A large area of shallow water beach and wetlands formed at the sea-river crossing site, which has become an important 'transfer station' and wintering and breeding place for bird migration at the inland Northeast Asia and Western Pacific Ring (Jiang 1999). Its wetlands have rich biodiversity and high productivity. It provides a variety of raw materials for human food, medicine, energy, and industry and provides crucial water storage, groundwater recharge, regional climate regulation, pollutant adsorption and filter, and disaster buffering (Zhang et al. 2010).

However, as a result of oil field development, increased human activity, and environmental change, the wetlands at the YRD region are experiencing regional degradation. Accordingly, phytocoenosium changes greatly, and the succession of plants presents specific features. To approach the mechanism of wetland degradation, biodiversity changes, and plant community succession, we performed a vegetation survey and a soil property analysis of the YRD region.

18.2 Methods

18.2.1 Test Area Description

The study area is located in Dongying, Shandong Province, in the northeastern part of the YRD region. It has a warm temperate continental climate, with an annual average temperature of 12.3 °C, an extreme maximal temperature of 41.9 °C, an extreme minimal temperature of −23.3 °C, an above 0 °C accumulated temperature of 4,783.5 °C, an above 10 °C accumulated temperature of 4,183 °C, an annual total solar radiation of 5,146–5,411 J m^{-2}, annual sunshine of 2,571–2,865 h with an average of 2,682 h (it is one of the most abundant regions of sunshine in China), an average frost-free period of 210 days, and annual precipitation of 574.4 mm, of which approximately 63.9 % is concentrated in summer. The annual evaporation is 1,962.1 mm (Zhang 2005), which is 3.6 times the rainfall, and spring is a period of intense evaporation, accounting for 51.7 % of the whole year. The study area is deposited with loess parent material on parent marine material. The mechanical component is mainly silt mixed with sand and sticky soil, and the soil has complex level changes. In this region, long long ago it was sea. Due to silting-up by Yellow river, land is formed. Thus soil formation lasted long time, and in different depth of soil, the mechanical component is various. Because the soil has a relatively short development time, it does not form a good structure. The soil pH is 6.79–8.87, with an average of 7.94. The underground water level is approximately 1.5 m, and the water salinity is higher.

18.2.2 Sites Setting

To investigate the wetland degradation process at the YRD region, the spatial difference of wetland degradation levels was used to study the temporal sequence of wetland degradation. According to the different types of halophytes, and considering the major factors that impact soil salinity changes, such as distance from the sea, altitude, and microtopographic changes, two typical sample lands of vegetation change were chosen from the study area. First, at the direction vertical to the coastline, different successional stages of wetland sample areas with typical cross-sectional structures were selected. According to the distance from the sea, the area comprises mainly bare land, *Suaeda glauca* community, Tamarix (*Tamarix chinensis*) community, and Imperata community. Second, the reed (*Phragmites australis*) community, with a wide range of ecological adaptations to the Yellow River, was chosen. According to the steady state, environment, and location of community, the communities were divided into three types: (a) growing well with higher community stability; (b) growing normally with potential community instability; and (c) growing poorly with degrading community.

18.2.3 Field Survey and Diversity Analysis

According to the quadrant survey method (Zhang 2005), sampling sites were set up at typical areas. Herbs were in 1 m × 1 m quadrants, eight repeats for each; while shrubs were in 10 m × 10 m quadrants, five repeats for each. Vegetation coverage, species, and others were surveyed for these samples. Vegetation coverage was estimated according to experience. Soil cross-sections were dug along the observation surface of sample sites, and soil samples were collected to measure the content of soil organic matter, total nitrogen, alkaline hydrolysis nitrogen, available phosphorus, available potassium, and soil salinity.

The indicators of biodiversity were as follows:

Species abundance index
Patrick Index D = S
D is the species available index
S indicates the number of species within the study area

$$\text{Importance value} = (\text{relative height} + \text{relative coverage} + \text{relative frequency})\,/\,3$$

Relative height was calculated as follows:

Relative height = average height of a species measured in field/sum of average heights for all species within a quadrant.

Relative coverage and relative frequency was calculated similarly.

$$\text{Hi} = -\sum \text{Pi}\ \ln\ \text{Pi}$$

Hi represents the species diversity index, whereas Pi is replaced by the importance value.

18.2.4 Soil Samples Collection and Pretreatment

Soils were collected by a multiple-point sampling method. Each test site had three evenly distributed sample collection plots, where soil cross-sectional samples of 0–20 and 20–40 cm were dug. Soil samples were packaged in sealed plastic bags, brought back to the laboratory, air dried, ground with a mortar, screened over a 1-mm sieve, and stored in sealed bags for analysis. Experimental data are the average values of analysis results from multiple samples at multiple points.

18.2.5 Determination of Soil Properties

Soil bulk density, noncapillary porosity, capillary porosity, total porosity, and other physical properties were determined by the cutting ring method (CACCSS 1983), and calculated according to the following formula:

1. Soil bulk density (g cm^{-3}) = dry soil weight within the cutting ring/volume of the cutting ring
2. Soil capillary porosity was calculated using the following formula:

$$p_c = (b - a - x)/v \times 100\%$$

   where p_c is the capillary porosity (%), b is the cutting ring weight with soil after 2–3 h of water absorption, a is the cutting ring weight (g), x is the dry soil weight within the cutting ring (g), and v is the volume of the cutting ring (cm^3).
3. Total soil porosity was calculated using the following formula:

$$p_t = (c - a - x)/v \times 100\%$$

   where p_t is the total soil porosity (%), c is the cutting ring weight with soil after 6 h of water absorption (g), a is the cutting ring weight (g), x is the dry soil weight within the cutting ring (g), and v is the volume of the cutting ring (cm^3).
4. Noncapillary porosity of soil was calculated using the following formula:

$$p_n = p_t - p_c$$

   where p_n is the soil's noncapillary porosity (%).

18.2.6 Determination of Soil Nutrients

Organic matter was measured by a potassium dichromate volumetric method, total nitrogen was measured by Kjeldahl distillation, alkaline hydrolysis nitrogen was measured by the alkaline hydrolysis diffusion method, available potassium was measured by ammonium acetate flame photometry, and available phosphorus was measured by the Olsen method (CACCSS 1983).

18.2.7 Determination of Soil Enzyme Activity

Urease activity was measured by colorimetry, catalase activity was measured by potassium permanganate titration, and peroxidase activity was measured by pyrogallol assay (Tabatabai 1994).

18.3 Results

18.3.1 Flora Features at Different Successional Stages of Wetlands

Vegetation was surveyed on test lands, and the results are shown in Tables 18.1 and 18.2.

As shown by the previous results, the vegetation at the YRD region has the following characteristics:

1. The type of vegetation is sparse, and the composition of fauna is poor, mainly dominated by Chenopodiaceae, Asteraceae, Poaceae, and Tamaricaceae. The salt-tolerant herbs are in dominance (Zhang et al. 2002).
2. Vegetation type is monotonous, and most are the monotonous halophyte, mainly including the *Suaeda glauca* community, the Tamarix community, the Imperata communities, and the reed community (Chen et al. 1992; Zhao 1995).
3. The community structure is simple, and most have no obvious layering. There are only one or two dominant species in each square meter of the quadrants.

18.3.2 Life Form Structures in Different Successional Stages

At the Imperata community stage, the community is mainly composed of perennial herbs. The importance value of constructive species is 35.8. The importance values of perennial herbs, annual herbs, and semi-shrubs are successively lower. At the Tamarix community stage, the importance value of constructive species is 36.23, and the importance value of the perennial reed is 17.03. In addition, annual herbaceous Limonium (*Limonium sinense*) appears, and is usually scattered within the Tamarix shrub. At the *Suaeda glauca* community stage, the importance value of

Table 18.1 Basic characteristics at different successional stages in wetlands

	Constructive species	Associated species	Distance to the sea (km)
Bare land	–	Sporadic *Suaeda glauca*	6
Suaeda glauca community	*Suaeda glauca*	*Tamarix*, reed	8
Tamarix community	*Tamarix*	*Suaeda glauca*, reed, Limonium, Cynanchum (*Cynanchum chinense*)	11.5
Imperata community	*Imperata*	*Ixeris chinensis*, reed, madder (*Phragmites australis*), Cephalanoplos (*Cirsium setosum*), Apocynum (*Apocynum venetum*), Salsola	16

Table 18.2 Plant community characteristics at different successional stages in wetlands

	Species	Relative height (m)	Relative coverage (%)	Relative frequency (%)	Importance value (%)
Suaeda glauca community	*Suaeda glauca*	34.4	95.4	66.6	65.47
	Tamarix	23.0	4.29	16.7	14.47
	Reed	42.6	0.4	16.7	19.90
Tamarix community	*Tamarix*	28.7	51.4	0.29	36.23
	Suaeda glauca	19.9	42.8	28.6	30.43
	Reed	21.7	4.4	25	17.03
	Cynanchum	25.5	1.2	14.3	13.67
	Limonium	4.2	0.2	3.5	2.63
Imperata community	*Imperata*	10.9	80.3	16.3	35.8
	Ixeris chinensis	10.7	7.25	14.0	10.7
	Reed	11.0	4.02	16.3	10.4
	Small reed	11.7	4.4	11.6	9.2
	Madder	9.4	2.87	9.3	7.2
	Apocynum	7.0	1.15	9.3	5.8
	Artemisia stelleriana	9.6	0.96	4.7	5.1
	Wild soybean	5.1	5.7	2.3	4.4
	Setaria viridis	9.0	0.95	2.3	4.1
	Cynanchum	5.9	0.76	4.7	3.8
	Aeluropus (*Aeluropus sinensis*)	3.4	0.76	7.0	3.7
	Chenopodium album	5.9	0.38	2.3	2.9

constructive species is 65.47. One species of shrub and one species of perennial herb are the community, and the ratio of importance value is 4.53:1.37:1. Bare land has the highest salinity with little vegetation, and only sporadic annual *Suaeda glauca* can be found. These results show that with increased degradation of wetlands, the importance value of constructive species gradually increases, and the life form structure of the plant community becomes simple.

18.3.3 Species Composition and Plant Diversity at Different Successional Stages

The surveyed quadrants from the wetlands in different community successional stages contained a total of 16 species of higher plants that belong to 9 families and 14 genera (Chen et al. 1992). Except for Gramineae, Chenopodiaceae, Asteraceae, and Papilionaceae, all are single family and single genera, such as Tamaricaceae, Apocynaceae, and Asclepiadaceae, reflecting the characteristics that saline-alkali land has a simple vegetation type and a mostly monotonous halophyte.

Plant diversity and community stability are closely related. As shown in Tables 18.1 and 18.2, the abundance of species and the diversity of community are consistent. Both abundance of species and diversity of community gradually decrease in the Imperata community, the Tamarix community, the *Suaeda glauca* community, and bare land. From bare land to Imperata community land, the species diversity basically shows a single peak variation.

18.3.4 Sequence of Vegetation Change During Wetland Degradation

18.3.4.1 Imperata Stage

At the Imperata stage, Imperata is the constructive species, and the major associated species are *Ixeris chinensis*, reed, madder (*Rubia cordifolia* L.), Cephalanoplos (*Cirsium setosum*), and Apocynum (*Apocynum venetum* L.). The habitat has a higher degree of soil desalination, often forming a single community. Vegetation coverage is over 80 %, and with the presence of 12 plant species. Human disturbances at this stage, such as overgrazing and extensive land reclamation, ground vegetation, increase the surface evaporation, and accelerate the accumulation rate of soil salinity to the ground surface. Thus, the surface soil salinity gradually increases, and plants with low salt tolerance gradually die in this region. This forms a vicious cycle, causing reverse vegetation succession and increased salinity, and forming a salt land meadow of Tamarix or *Suaeda glauca*.

18.3.4.2 Tamarix Stage

At the Tamarix stage, Tamarix is the constructive species, and the major associated species are *Suaeda glauca* and reed. The soil is mostly coastal saline soil, and vegetation coverage is approximately 40 %. Five plant species exist in this land type (Mao et al. 2003). If seawater intrudes or groundwater level increases, the ecological conditions rapidly deteriorate, causing the number and distribution of Tamarix populations to reduce and the probability of random extinction to increase; thus, plant community structure tends to be simpler. At low-altitude land with water accumulation, the community evolves to more salt-tolerant reed. At wetlands with seasonal water accumulation, the community evolves to *Suaeda glauca* wet meadow or bare land, and the developed soil becomes coastal meadow saline soil and coastal marsh saline soil (see Fig. 18.1).

Fig. 18.1 *Tamarix* stage (Photograph taken by Zhang Jianfeng)

18.3.4.3 *Suaeda glauca* Stage

At the *Suaeda glauca* stage, *Suaeda glauca* is the constructive species, and the major associated species are Tamarix and reed (see Fig. 18.2). The plant species composition is relatively monotonous, with only two to three types of plants. Soil is coastal meadow saline soil. The wetlands are highly unstable at this stage, and easily degrade to bare land.

18.3.4.4 Bare Land

Bare land is closer to the sea and often suffers from seawater invasion and drainage. This intertidal saline soil has a high salt content, and the meadow forming process is weak with only scattered *Suaeda glauca*.

18.3.5 Wetland Soil Characteristics at Different Succession Stages

Nutrient characteristics at different successional stages of wetlands were analyzed.

Soil samples were collected at different locations and measured for some parameters. The results are shown in Table 18.3.

As shown in Table 18.3, during wetland succession, the organic matter content in bare land, the *Suaeda glauca* community, and the Tamarix community, respectively,

Fig. 18.2 Growth of *Suaeda glauca* (Photograph taken by Zhang Jianfeng)

Table 18.3 Soil nutrients at different successional stages in wetlands

	Soil (cm)	Organic matter (g/kg)	Total nitrogen (g/kg)	Available nitrogen (mg/kg)	Available phosphorus (mg/kg)	Available potassium (mg/kg)
Bare land	0–20	2.4	0.21	10.83	8.29	65.98
	20–40	2.14	0.15	8.29	8.08	53.27
	Average	2.27	0.18	9.56	8.18	59.62
Suaeda glauca community	0–20	4.48	0.25	11.86	9.15	82.17
	20–40	3.29	0.17	12.94	9.58	132.68
	Average	3.89	0.21	12.4	9.36	107.43
Tamarix community	0–20	5.93	0.334	15.09	9.37	124.38
	20–40	3.73	0.271	12.94	10.01	74.56
	Average	4.83	0.30	14.02	9.69	99.47
Imperata community	0–20	15.9	0.747	41.63	10.66	138.13
	20–40	7.96	0.267	20.48	9.67	52.64
	Average	11.93	0.507	31.05	10.16	95.38

was reduced by 81.0 %, 67.4 %, and 59.5 %, compared with the Imperata community, and the total nitrogen content was reduced by 61.1 %, 59.6 %, and 41.2 %, respectively. Multiple comparison results show that the organic matter content difference between the Imperata community and the Tamarix community, the *Suaeda glauca* community, and bare land reaches an extremely significant level ($p < 0.01$); the difference between bare land and the Tamarix and the *Suaeda glauca* communities is significantly different ($p < 0.01$); and the difference between the Tamarix community and the *Suaeda glauca* community is not significant. The total nitrogen content difference between bare land and the Suaeda glauca community is not significant; the difference between bare land, the *Suaeda glauca* community, and the Tamarix community reaches a significant level ($p < 0.05$), and the difference between them and the Imperata community is extremely significant ($p < 0.01$).

Compared with the Imperata community, the nitrogen content in bare land, the *Suaeda glauca* community, and the Tamarix community decreased by 69.2 %, 60.1 %, and 53.5 %, respectively, and the available phosphorus decreased by 19.5 %, 7.9 %, and 4.7 %, respectively. The alkali-hydrolysis nitrogen content difference between bare land, the *Suaeda glauca* community, the Tamarix community, and the Imperata community is significant ($p < 0.01$), and the alkali-hydrolysis nitrogen content difference with bare land, *Suaeda glauca* community, and the Tamarix community is also significant ($p < 0.05$). The available phosphorus content from the Imperata community to bare land has not yet reached significant levels; however, with the succession of vegetation communities, available phosphorus content is gradually decreasing. The available potassium content does not show regularity with community succession, which may be the impact of the parent material of soil.

With the succession of communities in wetlands, from the Imperata community to bare land, the main nutrients of soil, including organic matter, total nitrogen, and available nitrogen, have an obvious decreasing trend. The differences of major nutrient parameters between bare land and the *Suaeda glauca*, Tamarix, and Imperata communities are all extremely significant, indicating that the successional bare land from the Imperata community at the wetland is an adverse succession. With the declination of the succession level, soil nutrient levels gradually decrease.

18.3.6 Soil Physical Properties at Different Successional Stages of Wetlands

The results of measurements of the physical properties of soil are shown in Table 18.4.

As shown in the table, from bare land to the Imperata community, soil bulk density decreases, and the total porosity and capillary porosity increase, indicating that, as the salinity degree increases, soil physical properties gradually deteriorate. Reduced soil physical properties, enhanced soil bulk density, decreased porosity, and more compact soil are not conducive to soil desalinization.

18.3.7 Soil Enzyme Activity at Different Successional Stages of Wetlands

The results of the soil enzyme activity are shown in Table 18.5. As shown in the table, at the vertical direction, the levels of urease and catalase activity are higher

Table 18.4 Soil physical properties at different successional stages in wetlands

Plot type	Soil layer (cm)	Bulk density (g/cm^3)	Total porosity (%)	Capillary porosity (%)	Noncapillary porosity (%)
Bare land	0–20	1.83	32.7	31.4	1.3
	20–40	1.7	43.2	41.5	1.7
	Average	1.765	38.0	36.45	1.5
Suaeda glauca community	0–20	1.67	40.3	38.4	1.90
	20–40	1.54	42.1	40.3	1.8
	Average	1.605	41.2	39.3	1.85
Tamarix community	0–20	1.55	43.3	40.5	2.8
	20–40	1.65	41.5	39.8	1.7
	Average	1.6	42.4	40.1	2.2
Imperata community	0–20	1.5	44.9	41.7	3.2
	20–40	1.54	42.0	40.2	1.8
	Average	1.52	43.45	40.95	2.5

Table 18.5 Soil enzyme activity at different successional stages in wetlands

Plot type	Soil layer (cm)	Urease NH_3-N/mg.g^{-1}	Catalase 0.1 mol/l $KMnO_4$/ml.g^{-1}	Peroxidase purple gallic element/mg.g^{-1}
Bare land	0–20	0.023	0.63	1.11
	20–40	0.013	0.55	1.23
	Average	0.018	0.59	1.17
Suaeda glauca community	0–20	0.079	0.85	0.917
	20–40	0.049	0.60	0.997
	Average	0.064	0.725	0.897
Tamarix community	0–20	0.098	0.87	0.977
	20–40	0.057	0.60	0.997
	Average	0.077	0.735	0.987
Imperata community	0–20	0.349	1.00	0.707
	20–40	0.213	0.75	0.759
	Average	0.281	0.875	0.733

at the upper layer than at the lower layer. However, the level of peroxidase is the opposite, the upper being lower than the lower, which is inconsistent with the soil peroxidase activity in woodlands. Changes in soil enzyme activity may be related to the differences in soil structure.

At different successional stages from the Imperata community to bare land, the enzyme activities of urease and catalase decrease. However, the peroxidase activity is not regular, with the highest level at bare land and the lowest level at the Imperata community, which may be related to the soil litter fall and seasonal water accumulation at wetlands.

18.4 Discussion and Conclusions

The evolution of vegetation and the degradation of wetlands occur nearly simultaneously. Soil salinity accelerates the reverse succession of plant communities, which in turn transforms the physical and chemical properties of soil in a negative direction for plant growth (Han et al. 2000; Li et al. 2001). The analysis of physical and chemical properties of soil at different community successional stages reveals that, along with the degradation of vegetation, the physical and chemical properties of soil change correspondingly and the salinity level of soil gradually increases, making it non-conducive to the survival of the Imperata community.

At the Imperata community stage, the soil is slightly alkalized, and the surface salinity is generally less than 0.3 %. Under such conditions, Imperata can grow well, soil structure is good, soil bulk density is low, porosity is high, soil is more fertile, and the organic matter content is high, all of which are conducive to plant growth. When the vegetation is not destroyed, the soil remains at this stage or improves further.

As the salinity increases, the soil gradually becomes unsuitable for Imperata growth, and some halophytes begin to appear, forming the halophyte community stage. At this stage, soil salinity is aggravated, the surface salinity increases to above 0.8 %, soil structure becomes more compact and solid, soil bulk density increases, and the porosity is low.

When vegetation degrades to the *Suaeda glauca* community, the soil salinity level further increases, soil evolves into severe saline-alkali soil, soil salinity elevates significantly, the degree of salinization aggravates, soil texture becomes tight, permeability is poor, and soil fertility declines.

Further degradation of wetlands causes vegetation disappearance, leaving bare land with only sporadic *Suaeda glauca*. The soil environment is extremely harsh, the salt content is high, and soil becomes compact and sterile, with only 2.4 g/kg of organic matter content. In such an environment, it is difficult for a plant to grow.

The water–salt movement of soil is the core of saline soil evolution and an important factor affecting wetland vegetation (Zhang and Jiang 2010). Under certain conditions, the water–salt movement of soil is the driving force for plant community succession (Song et al. 2003). Salt accumulation in soil is mainly limited by soil permeability and the underground water level, whereas soil permeability is correlated with the structure and texture of soil. The Imperata community has better soil structure with good porosity, and soil water has a high infiltration rate and amount. Salt accompanies water, which necessarily makes salt leach and migrate downward, resulting in reduced soil salinization and a well-developed halophyte community.

Bare land → *Suaeda glauca* → Tamarix → Imperata succession is a process of soil desalination and also a forward successional process. If the soil desalination process is relatively smooth, soil will develop into an Imperata or Aeluropus community; if there happens to be fresh water accumulation, a reed community will appear, and it will take a reverse succession to form a Tamarix-*Suaeda glauca* community.

During the soil's evolution, soil's organic matter has an important influence on the physical and chemical properties of soil. It can change soil structure with the role of desalination and salt suppression. During the adverse succession of community, as the soil organic matter content declines, soil salinity will increase. The soil structure becomes tighter with the reduction of organic matter, the bulk density increases, and the porosity is reduced. Thus, increasing soil organic matter content is an effective way to improve saline-alkali soil.

In the natural state, the change of community diversity depends on community structure and function, and also reflects the dynamic changes of the physical and chemical properties of the soil during its evolution (Glenn-Lewin et al. 1992; Ren et al. 2001; Baoyin et al. 2003). With the degradation of vegetation, the organic matter content of the soil decreases, and the soil becomes poorer and poorer, resulting in a decline in species diversity. The wetland degradation at the YRD region is closely related to the fragile ecological environment in this area, the different understanding of natural processes by humans at different periods, and the intensive development of water resources (Xing et al. 2005). The vegetation growth (coverage and species) and changes in soil properties could be the direct indicators of land degradation (Liu 1992; Zhang 1995).

References

Baoyin TGT, Liu ML, Li XL (2003) The study on dynamics succession of community in degenerated. Acta Phytoecol Sin 2:270–277

Chen HB, Zheng YJ, Li FZ (1992) Shandong flora (Volume 1, Volume 2). Qingdao Publishing House, Qingdao

Committee of Agricultural Chemistry of Chinese Soil Society (1983) Conventional analysis methods in soil agricultural chemistry. Science Press, Beijing

Glenn-Lewin D, Peet R, Veblen T (1992) Succession: theory and prediction. Chapman & Hall, London, pp 180–190

Han YZ, Tian LY, Xu XG (2000) Yellow River Delta region wetland ecosystems and preliminary study of their protection. Environ Sci Technol 2:10–13

Jiang ZH (1999) Forest ecological engineering projects and the sustainable development at Yellow River Delta region. Forest Res 5:447–451

Li WL, Li ZZ, Zhou ZX (2001) A quantitative research on artificial plant community. J Lanzhou Univ (Nat Sci) 5:98–104

Liu JG (1992) Advance in modern ecology. China Science and Technology Press, Beijing, pp 69–76

Mao HY, Zhao QJ, Gao Q (2003) Ideas and models of natural resource development under ecological constrains. Nat Sci Press 4:459–466

Ren H, Cai XA, Rao XQ et al (2001) The theory on succession of plant community. Ecol Sci 4:59–67

Song YM, Zhang JF, Xing SJ et al (2003) Flora feature and vegetation recovery technology at heavy saline-alkali soil of Yellow River Delta region. J Northeast Forest Univ 6:87–90

Tabatabai MA (1994) Soil enzymes. In: Weaves RW, Angel GS, Bottomley PS (eds) Methods of soil analysis. Part 2, Microbiological and biochemical properties, vol 5, SSSA book series. Soil Science Society of America Journal, Madison

Xing SJ, Zhang JF, Song YM et al (2005) Ecological functions and ecological restoration techniques at wetlands of Yellow River Delta region. Shandong Forest Sci Technol 2:24–27

Xu XG (1996) Assessment and early warning study at ecosystem of Yellow River Delta region. Acta Ecol Sin 5:461–468

Zhang JT (1995) Methods in quantitative vegetation ecology. China Science and Technology Press, Beijing, pp 97–255

Zhang JF (2005) Causes of wetland degradation and ecological restoration in Yellow River Delta region. Forest Study China 2:15–18

Zhang JF, Jiang JM (2010) Soil salinization and ecological remediation by planting trees in China. In: Proceedings of 2010 international conference on mechanic automation and control engineering. Wuhan, China, pp 1349–1352

Zhang JF, Xing SJ, Xi JB et al (2002) The environmental issues faced by sustainable development at Yellow River Delta region and the forestry development strategies. J Northeast Forest Univ 6:115–119

Zhang JF, Chen GC, Xing SJ et al (2010) Carbon sequestration of black locust forests in the Yellow River Delta region, China. Int J Sustain Dev World Ecol 6:475–480

Zhang JF, Chen GC, Xing SJ (2011) Water shortages and countermeasures for sustainable utilisation in the context of climate change in the Yellow River Delta region, China. Int J Sustain Dev World Ecol 2:177–185

Zhao YM (1995) Scientific survey of Yellow River Delta region natural reserve. China Forestry Press, Beijing

Index

A

Afforestation, 34–39, 110, 112, 114, 146, 159
Agroforestry, 41–53, 71
Amelioration, 4, 7, 12, 15–21, 39, 49, 109–115, 142

B

Biodiversity, 53, 78, 89, 90, 98, 101, 166, 168
Bioenergy resources, 80, 158
Biomass energy, 75–81, 157–164
  resources, 157–164
Biomass plantation, 77
Biomass production, 67–72, 75–81, 163, 164
Black locust, 62, 63, 71, 145–154, 157–164
Black locust forest, 149, 154, 157–164

C

Characteristics of salt-affected soil, 4, 15–21
Coastal areas, 24, 76–78, 80, 98
Coastal ecosystem, 78, 98
Coastline, 56, 76, 86, 97, 109, 135, 147, 159, 167
Contamination, 9, 21, 23, 24, 27, 76, 91

D

Decreasing recharge, 18, 21, 62
Deforestation, 43
Desertification, 9, 23, 43, 56, 68, 86, 97, 109, 135, 159

E

East China Sea, 75–81
Environmental problem, 23, 44, 53, 56, 68, 77, 86, 96, 160

F

Fertilizers, 6, 37, 45, 60–61, 64, 91, 105, 151, 152
Forage crops, 59, 151
Forestation, 3, 7, 12, 33–39, 43, 110, 112, 114, 122, 146, 158, 164

G

Groundwater recharge, 16, 18, 27, 59, 62, 102, 166

I

Impact, 10, 24, 26, 27, 93–105, 142, 163, 167, 175
Increasing discharge, 15, 18–19, 21, 62
Intercropping, 36, 42, 51–53, 63, 71

L

Land runoff, 76
Land use, 3, 5, 10, 11, 24, 25, 43, 47, 49, 57, 62, 71, 89, 91, 96, 97, 111, 126, 149–152, 158, 159, 163
Leguminous crops, 47

M

Maintaining biodiversity, 78
Manure, 36, 60–61, 63
Marginal land, 75
Mulch, 28, 44, 49, 61
Multipurpose trees, 49, 50

N

Nitraria sibirica, 117–122, 128, 129

J. Zhang, *Coastal Saline Soil Rehabilitation and Utilization Based on Forestry Approaches in China*, DOI 10.1007/978-3-642-39915-2,

O
Organic matter (OM), 18, 20, 36, 46, 49, 60, 76, 89, 114, 128, 138, 140–142, 147, 148, 151, 153, 154, 157, 164, 168, 169, 173–175, 177, 178

P
Physiological indicator, 129, 138
Phytocoenosium, 166
Plantation, 13, 37, 39, 42, 43, 50, 77, 79, 145–154, 160
Planting techniques, 33–39, 110, 111, 117–119
Population growth, 3, 11, 24, 41–44, 76, 79, 157
Power plants, 68, 72
Provenance, 120–122

R
Renewable energy, 79, 156, 158, 164
Riverbank, 135–142, 146, 148, 149, 151, 153, 154
Rural development, 42

S
Saline soil distribution, 11–13
Saline soil resources, 68–69
Salinity stress, 110, 126, 128, 131
Salinity tolerance, 21, 33–39, 50, 126, 129, 131, 132, 138
Salinity tolerance of trees, 34, 126, 138
Salinization, 5, 7, 11–13, 18, 21, 23–28, 36, 42, 44–47, 53, 56–60, 63, 68, 69, 71, 86, 89, 97, 98, 109, 110, 125, 135, 136, 141, 142, 150, 151, 159, 165, 175, 177
Salt-affected soil, 3–7, 9–13, 15–21, 27, 33, 34, 57, 117–122, 150, 159
Salt concentration, 35, 49, 53, 59, 60, 109, 110, 112–114, 127, 141, 149
Salt tolerance, 17, 39, 61, 62, 110–112, 118, 119, 172
Salt-tolerant grass, 61, 62
Seed germination, 110, 112, 114, 121
Seedling nursing, 37, 126, 127
Shelterbelt, 43, 48, 50–51, 62, 71, 104
Site preparation, 36, 37, 39, 111–115, 117, 119–122
Site selection, 37
Socio-economic development, 3, 9–11, 24, 36, 39, 44, 55, 76, 79, 89, 96, 98, 100, 102, 103, 110, 146, 158
Soil amelioration, 49, 142
Soil enzymes activity, 149, 152–154, 169–170, 176–177
Soil erosion, 7, 12, 42, 44, 50, 135–142, 146, 153, 154
Soil improvement, 21, 71, 111, 151
Soil salinazation, 5, 9, 18, 24, 26, 53, 58, 89, 110, 125, 136, 141, 150, 177
Soil salinity, 5, 7, 11, 12, 15, 16, 25, 26, 36, 37, 47, 50, 52, 53, 61, 62, 64, 69, 89, 95, 121, 125–132, 136, 138–140, 142, 147–150, 153, 154, 167, 168, 172, 177, 178
Surface drainage, 5, 20, 25, 27, 59, 60
Sustainable utilization, 75

T
Tamarix chinensis, 90, 109–115, 128, 129, 131, 132, 137, 138, 140, 167
Tree species selection, 35–36, 110, 117, 125–132

U
Urban salinization, 23–28

W
Water shortage, 89, 93–105
Water wastage, 101–103
Wetland degradation, 85–91, 165–178
Wetland restoration, 90–91
White death, 6

Y
Yellow River Delta (YRD), 39, 52, 55–64, 86–91, 93, 95–101, 103, 104, 109–111, 115, 117, 118, 125, 126, 128, 135, 136, 138, 142, 146, 149, 151, 152, 154, 159, 160, 163, 165–167, 170, 178

MIX
Papier aus verantwortungsvollen Quellen
Paper from responsible sources
FSC® C105338

If you have any concerns about our products,
you can contact us on
ProductSafety@springernature.com

In case Publisher is established outside the EU,
the EU authorized representative is:
Springer Nature Customer Service Center GmbH
Europaplatz 3, 69115 Heidelberg, Germany

Printed by Libri Plureos GmbH
in Hamburg, Germany